Multimodal Foundation Models: From Specialists to General-Purpose Assistants

Other titles in Foundations and Trends® in Computer Graphics and Vision

Computational Imaging Through Atmospheric Turbulence
Stanley H. Chan and Nicholas Chimitt
ISBN: 978-1-63828-999-9

Towards Better User Studies in Computer Graphics and Vision
Zoya Bylinskii, Laura Herman, Aaron Hertzmann, Stefanie Hutka and Yile Zhang
ISBN: 978-1-63828-172-6

An Introduction to Neural Data Compression
Yibo Yang, Stephan Mandt and Lucas Theis
ISBN: 978-1-63828-174-0

Learning-based Visual Compression
Ruolei Ji and Lina J. Karam
ISBN: 978-1-63828-112-2

Vision-Language Pre-Training: Basics, Recent Advances, and Future Trends
Zhe Gan, Linjie Li, Chunyuan Li, Lijuan Wang, Zicheng Liu and Jianfeng Gao
ISBN: 978-1-63828-132-0

Semantic Image Segmentation: Two Decades of Research
Gabriela Csurka, Riccardo Volpi and Boris Chidlovskii
ISBN: 978-1-63828-076-7

Multimodal Foundation Models: From Specialists to General-Purpose Assistants

Chunyuan Li
Microsoft Corporation
chunyl@microsoft.com

Zhe Gan
Microsoft Corporation
zhgan@microsoft.com

Zhengyuan Yang
Microsoft Corporation
zhengyang@microsoft.com

Jianwei Yang
Microsoft Corporation
jianwyan@microsoft.com

Linjie Li
Microsoft Corporation
linjli@microsoft.com

Lijuan Wang
Microsoft Corporation
lijuanw@microsoft.com

Jianfeng Gao
Microsoft Corporation
jfgao@microsoft.com

now
the essence of knowledge
Boston — Delft

Foundations and Trends® in Computer Graphics and Vision

Published, sold and distributed by:
now Publishers Inc.
PO Box 1024
Hanover, MA 02339
United States
Tel. +1-781-985-4510
www.nowpublishers.com
sales@nowpublishers.com

Outside North America:
now Publishers Inc.
PO Box 179
2600 AD Delft
The Netherlands
Tel. +31-6-51115274

The preferred citation for this publication is

C. Li *et al.*. *Multimodal Foundation Models: From Specialists to General-Purpose Assistants*. Foundations and Trends® in Computer Graphics and Vision, vol. 16, no. 1-2, pp. 1–214, 2024.

ISBN: 978-1-63828-336-2

Foundations and Trends® in Computer Graphics and Vision

Volume 16, Issue 1-2, 2024

Editorial Scope

Topics

Foundations and Trends® in Computer Graphics and Vision publishes survey and tutorial articles in the following topics:

- Rendering
- Shape
- Mesh simplification
- Animation
- Sensors and sensing
- Image restoration and enhancement
- Segmentation and grouping
- Feature detection and selection
- Color processing
- Texture analysis and synthesis
- Illumination and reflectance modeling
- Shape representation
- Tracking
- Calibration
- Structure from motion
- Motion estimation and registration
- Stereo matching and reconstruction
- 3D reconstruction and image-based modeling
- Learning and statistical methods
- Appearance-based matching
- Object and scene recognition
- Face detection and recognition
- Activity and gesture recognition
- Image and video retrieval
- Video analysis and event recognition
- Medical image analysis
- Robot localization and navigation

Information for Librarians

Foundations and Trends® in Computer Graphics and Vision, 2024, Volume 16, 4 issues. ISSN paper version 1572-2740. ISSN online version 1572-2759. Also available as a combined paper and online subscription.

Contents

Multimodal Foundation Models: From Specialists to General-Purpose Assistants

Chunyuan Li, Zhe Gan, Zhengyuan Yang, Jianwei Yang, Linjie Li, Lijuan Wang and Jianfeng Gao

Microsoft Corporation, USA; chunyl@microsoft.com, zhgan@microsoft.com, zhengyang@microsoft.com, jianwyan@microsoft.com, linjli@microsoft.com, lijuanw@microsoft.com, jfgao@microsoft.com

---

ABSTRACT

This monograph presents a comprehensive survey of the taxonomy and evolution of multimodal foundation models that demonstrate vision and vision-language capabilities, focusing on the transition from specialist models to general-purpose assistants. The research landscape encompasses five core topics, categorized into two classes. (*i*) We start with a survey of well-established research areas: multimodal foundation models pre-trained for specific purposes, including two topics – methods of learning vision backbones for visual understanding and text-to-image generation. (*ii*) Then, we present recent advances in exploratory, open research areas: multimodal foundation models that aim to play the role of general-purpose assistants, including three topics – unified vision models inspired by large language models (LLMs), end-to-end training of multimodal LLMs, and chaining multimodal tools with LLMs. The target audiences of the

Chunyuan Li, Zhe Gan, Zhengyuan Yang, Jianwei Yang, Linjie Li, Lijuan Wang and Jianfeng Gao (2024), "Multimodal Foundation Models: From Specialists to General-Purpose Assistants", Foundations and Trends® in Computer Graphics and Vision: Vol. 16, No. 1-2, pp 1–214. DOI: 10.1561/0600000110.

monograph are researchers, graduate students, and professionals in computer vision and vision-language multimodal communities who are eager to learn the basics and recent advances in multimodal foundation models.

1

Introduction

Vision is one of the primary channels for humans and many living creatures to perceive and interact with the world. One of the core aspirations in artificial intelligence (AI) is to develop AI agents to mimic such an ability to effectively perceive and generate visual signals, and thus reason over and interact with the visual world. Examples include recognition of the objects and actions in the scenes, and creation of sketches and pictures for communication. Building foundational models with visual capabilities is a prevalent research field striving to accomplish this objective.

In Figure 1.1, the evolution of AI is depicted, beginning with a specialized entity symbolized by a desktop computer and progressing to a versatile, general-purpose assistant represented by Doraemon. The choice of Doraemon serves a dual purpose: to highlight AI's capability to perform a broad spectrum of tasks and to emphasize its constant readiness to support human needs.

Over the last decade, the field of AI has experienced a fruitful trajectory in the development of models. We divide them into four categories, as illustrated in Figure 1.2. The categorization can be shared among different fields in AI, including language, vision and multimodality. We

Figure 1.1: The visual illustration of AI evolution from specialists to general-purpose assistants.

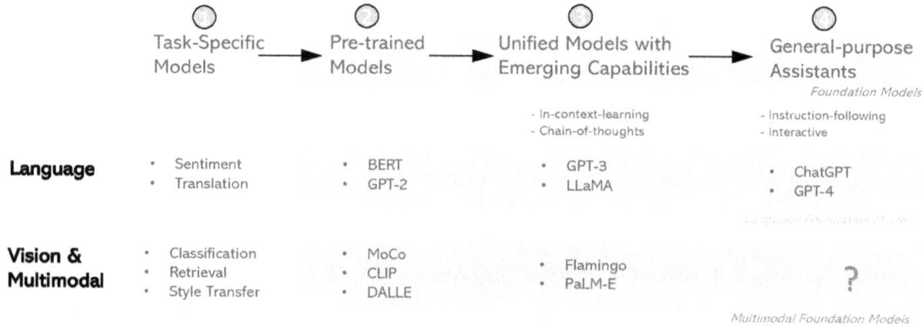

Figure 1.2: Illustration of foundation model development trajectory for language and vision/multi-modality. Among the four categories, the first category is the task-specific model, and the last three categories belong to foundation models, where these foundation models for language and vision are grouped in green and blue blocks, respectively. Some prominent properties of models in each category are highlighted. By comparing the models between language and vision, we are foreseeing that the transition of multimodal foundation models follows a similar trend: from the pre-trained model for specific purpose, to unified models and general-purpose assistants. However, research exploration is needed to figure out the best recipe, which is indicated as the question mark in the figure, as technical details of GPT-4V and Gemini [426] stay private.

first use language models in NLP to illustrate the evolution process. (*i*) At the early years, task-specific models are developed for individual datasets and tasks, typically being trained from scratch. (*ii*) With large-scale pre-training, language models achieve state-of-the-art performance on many established language understanding and generation tasks, such as BERT [98], RoBERTa [277], T5 [363], DeBERTa [167] and GPT-2 [362]). These pre-trained models serve the basis for downstream task adaptation. (*iii*) Exemplified by GPT-3 [37], large language models (LLMs) unify various language understanding and generation tasks into one model. With web-scale training and unification, some emerging capabilities appear, such as in-context-learning and chain-of-thoughts. (*iv*) With recent advances in human-AI alignment, LLMs start to play the role of general-purpose assistants to follow human intents to complete a wide range of language tasks in the wild, such as ChatGPT [333] and GPT-4 [334]. These assistants exhibit interesting capabilities, such as interaction and tool use, and lay a foundation for developing general-purpose AI agents. It is important to note that the latest iterations of foundation models build upon the noteworthy features of their earlier counterparts while also providing additional capabilities.

Inspired by the great successes of LLMs in NLP, it is natural for researchers in the computer vision and vision-language community to ask the question: what is the counterpart of ChatGPT/GPT-4 for vision, vision-language and multi-modal models? There is no doubt that vision pre-training and vision-language pre-training (VLP) have attracted a growing attention since the birth of BERT, and has become the mainstream learning paradigm for vision, with the promise to learn universal transferable visual and vision-language representations, or to generate highly plausible images. Arguably, they can be considered as the early generation of multimodal foundation models, just as BERT/GPT-2 to the language field. While the road-map to build general-purpose assistants for language such as ChatGPT is clear, it is becoming increasingly crucial for the research community to explore feasible solutions to building its counterpart for computer vision: the general-purpose visual assistants. Overall, building general-purpose agents has been a long-standing goal for AI. LLMs with emerging properties have significantly reduced the cost of building such agents for language tasks. Similarly, we foresee emerging capabilities from vision models, such as following the instructions composed by various visual prompts like user-uploaded images, human-drawn clicks, sketches and mask, in addition to text prompt. Such strong zero-shot visual task composition capabilities can significantly reduce the cost of building AI agents.

In this monograph, we limit the scope of multimodal foundation models to the vision and vision-language domains. Recent survey papers on related topics include (*i*) *image understanding models* such as self-supervised learning [193], [196], [339], segment anything (SAM) [552], [554], (*ii*) *image generation models* [553], [592], and (*iii*) *vision-language pre-training (VLP)*. Existing VLP survey papers cover VLP methods for task-specific VL problems before the era of pre-training, image-text tasks, core vision tasks, and/or video-text tasks [54], [109], [129], [231], [379], [551], [560]. Two recent survey papers cover the integration of vision models with LLM [16], [525].

Among them, [129] is a survey on VLP that covers the CVPR tutorial series on *Recent Advances in Vision-and-Language Research* in 2022 and before. This work summarizes the CVPR tutorial on *Recent Advances in Vision Foundation Models* in 2023. Different from the aforementioned survey papers that focus on literature review of a given research topic, this monograph presents our perspectives on the role transition of multimodal foundation models from specialists to general-purpose visual assistants, in the era of large language models. The contributions of this survey are summarized as follows.

- We provide a comprehensive and timely survey on modern multimodal foundation models, not only covering well-established models for visual representation learning and image generation, but also summarizing emerging topics for the past 6 months inspired by LLMs, including unified vision models, training and chaining with LLMs.
- The monograph is positioned to provide the audiences with the perspective to advocate a transition in developing multimodal foundation models. On top of great modeling successes for specific vision problems, we are moving towards building general-purpose assistants that can follow human intents to complete a wide range of computer vision tasks in the wild. We provide in-depth discussions on these advanced topics, demonstrating the potential of developing general-purpose visual assistants.

1.1 What are Multimodal Foundation Models?

As elucidated in the Stanford foundation model paper [35], AI has been undergoing a paradigm shift with the rise of models (*e.g.*, BERT, GPT family, CLIP [360] and DALL-E [367]) trained on broad data that can be adapted to a wide range of downstream tasks. They call these models *foundation models* to underscore their critically central yet incomplete character: homogenization of the methodologies across research communities and emergence of new capabilities. From a technical perspective, it is *transfer learning* that makes foundation models possible, and it is *scale* that makes them powerful. The emergence of foundation models has been predominantly observed in the NLP domain, with examples ranging from BERT to ChatGPT. This trend has gained traction in recent years, extending to computer vision and other fields. In NLP, the introduction of BERT in late 2018 is considered as the inception of the foundation model era. The remarkable success of BERT rapidly stimulates interest in self-supervised learning in the computer vision community, giving rise to models such as SimCLR [62], MoCo [163], BEiT [26], and MAE [162]. During the same time period, the success of pre-training also significantly promotes the vision-and-language multimodal field to an unprecedented level of attention.

In this monograph, *we focus on multimodal foundation models, which inherit all properties of foundation models discussed in the Stanford paper [35], but with an emphasis on models with the capability to deal with vision and vision-language modalities.* Among the ever-growing literature, we categorize multimodal foundation models in Figure 1.3, based on their functionality and generality. For each category, we present exemplary models that demonstrate the primary capabilities inherent to these multimodal foundation models.

- **Visual Understanding Models.** (Highlighted with orange in Figure 1.3) Learning general visual representations is essential to build vision foundation models, as pre-training a strong vision backbone is fundamental to all types of computer vision downstream tasks, ranging from image-level (*e.g.*, image classification, retrieval, and captioning), region-level (*e.g.*, detection and grounding) to pixel-level tasks (*e.g.*, segmentation). We group the methods into three categories, depending on the types of supervision signals used to train the models.

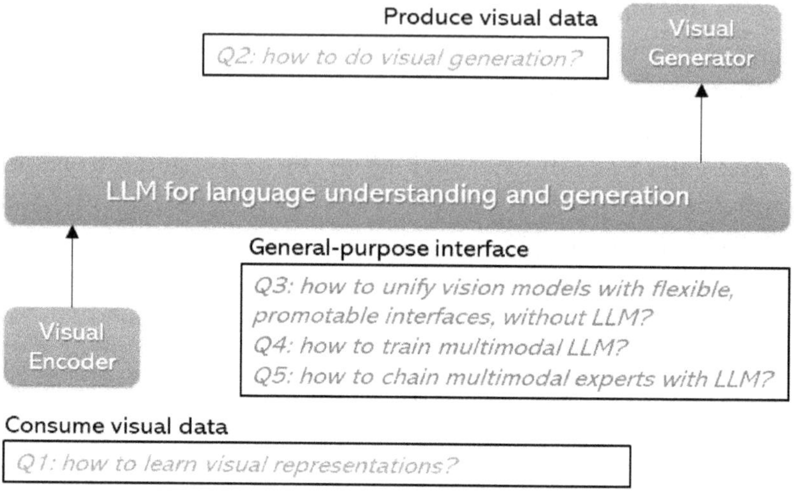

Figure 1.3: Illustration of three representative problems that multimodal foundation models aim to solve in this monograph: visual understanding tasks (orange), visual generation tasks (green), and general-purpose interface (blue) with language understanding and generation.

- **Label supervision.** Datasets like ImageNet [212] and ImageNet21K [374] have been popular for supervised learning, and larger-scale proprietary datasets are also used in industrial labs [406], [415], [548].
- **Language supervision.** Language is a richer form of supervision. Models like CLIP [360] and ALIGN [195] are pre-trained using a contrastive loss over millions or even billions of noisy image-text pairs mined from the Web. These models enable zero-shot image classification, and make traditional computer vision (CV) models to perform open-vocabulary CV tasks. We advocate the concept of *computer vision in the wild*,[1] and encourage the development and evaluation of future foundation models for this.
- **Image-only self-supervision.** This line of work aims to learn image representations from supervision signals mined from the images themselves, ranging from contrastive learning [62], [163],

[1] Computer-Vision-in-the-Wild Readings.

non-contrastive learning [46], [77], [144], to masked image modeling [26], [162].

  - **Multimodal fusion, region-level and pixel-level pre-training.** Besides the methods of pre-training image backbones, we will also discuss pre-training methods that allow multimodal fusion (*e.g.*, CoCa [530], Flamingo [4]), region-level and pixel-level image understanding, such as open-set object detection (*e.g.*, GLIP [241]) and promptable semgentation (*e.g.*, SAM [206]). These methods typically rely on a pre-trained image encoder or a pre-trained image-text encoder pair.

- **Visual Generation Models.** (Highlighted with green in Figure 1.3) Recently, foundation image generation models have been built, due to the emergence of large-scale image-text data. The techniques that make it possible include the vector-quantized VAE methods [370], diffusion-based models [99] and auto-regressive models.

  - **Text-conditioned visual generation.** This research area focuses on generating faithful visual content, including images, videos, and more, conditioned on open-ended text descriptions/prompts. Text-to-image generation develops generative models that synthesize images of high fidelity to follow the text prompt. Prominent examples include DALL-E [367], DALL-E 2 [366], Stable Diffusion [375], [412], Imagen [382], and Parti [531]. Building on the success of text-to-image generation models, text-to-video generation models generate videos based on text prompts, such as Imagen Video [172] and Make-A-Video [403].
  - **Human-aligned visual generator.** This research area focuses on improving the pre-trained visual generator to better follow human intentions. Efforts have been made to address various challenges inherent to base visual generators. These include improving spatial controllability [511], [562], ensuring better adherence to text prompts [31], supporting flexible text-based editing [36], and facilitating visual concept customization [380].

- **General-purpose Interface.** (Highlighted with blue in Figure 1.3) The aforementioned multimodal foundation models are designed for specific purposes – tackling a specific set of CV problems/tasks. Recently, we see an emergence of general-purpose models that lay the basis of AI agents. Existing efforts focus on three research topics. The first topic aims to unify models for visual understanding and generation. These models are inspired by the unification spirit of LLMs in NLP, but do not explicitly leverage pre-trained LLM in modeling. In contrast, the other two topics embrace and involve LLMs in modeling, including training and chaining with LLMs, respectively.

- **Unified vision models for understanding and generation.** In computer vision, several attempts have been made to build a general-purpose foundation model by combining the functionalities of specific-purpose multimodal models. To this end, a unified model architecture is adopted for various downstream computer vision and vision-language (VL) tasks. There are different levels of unification. First, a prevalent effort is to bridge vision and language by converting all closed-set vision tasks to open-set ones, such as CLIP [360], GLIP [242], OpenSeg [137], *etc.* Second, the unification of different VL understanding tasks across different granularity levels is also actively explored, such as I/O unification methods like UniTAB [508], Unified-IO [287]), Pix2Seq-v2 [65] and functional unification methods like GPV [151], GLIP-v2 [559]) and X-Decoder [599]. In the end, it is also necessitated to make the models more interactive and promptable like ChatGPT, and this has been recently studied in SAM [206] and SEEM [601].
- **Training with LLMs.** Similar to the behavior of LLMs, which can address a language task by following the instruction and processing examples of the task in their text prompt, it is desirable to develop a visual and text interface to steer the model towards solving a multimodal task. By extending the capability of LLMs to multimodal settings and training the model end-to-end, multimodal LLMs or large multimodal models are developed, including Flamingo [4] and Multimodal GPT-4 [334].
- **Chaining tools with LLM.** Exploiting the tool use capabilities of LLMs, an increasing number of studies integrate LLMs such as ChatGPT with various multimodal foundation models

to facilitate image understanding and generation through a conversation interface. This interdisciplinary approach combines the strengths of NLP and computer vision, enabling researchers to develop more robust and versatile AI systems that are capable of processing visual information and generating human-like responses via human-computer conversations. Representative works include Visual ChatGPT [477] and MM-ReAct [513].

1.2 Definition and Transition from Specialists to General-Purpose Assistants

Based on the model development history and taxonomy in NLP, we group multimodal foundation models in Figure 1.3 into two categories.

- **Specific-Purpose Pre-trained Vision Models** cover most existing multimodal foundation models, including visual understanding models (*e.g.*, CLIP [360], SimCLR [62], BEiT [26], SAM [206]) and visual generation models (*e.g.*, Stable Diffusion [375], [412]), as they present powerful transferable ability for specific vision problems.
- **General-Purpose Assistants** refer to AI agents that can follow human intents to complete various computer vision tasks in the wild. The meanings of general-purpose assistants are two-fold: (*i*) generalists with unified architectures that could complete tasks across different problem types, and (*ii*) easy to follow human instruction, rather than replacing humans. To this end, several research topics have been actively explored, including unified vision modeling [287], [559], [599], training and chaining with LLMs [264], [477], [513], [593].

1.3 Who Should Read this Monograph?

This monograph is based on our CVPR 2023 tutorial,[2] with researchers in the computer vision and vision-language multimodal communities as our primary target audience. It reviews the literature and explains topics to those who seek to learn the basics and recent advances in

[2]https://vlp-tutorial.github.io/2023/index.html

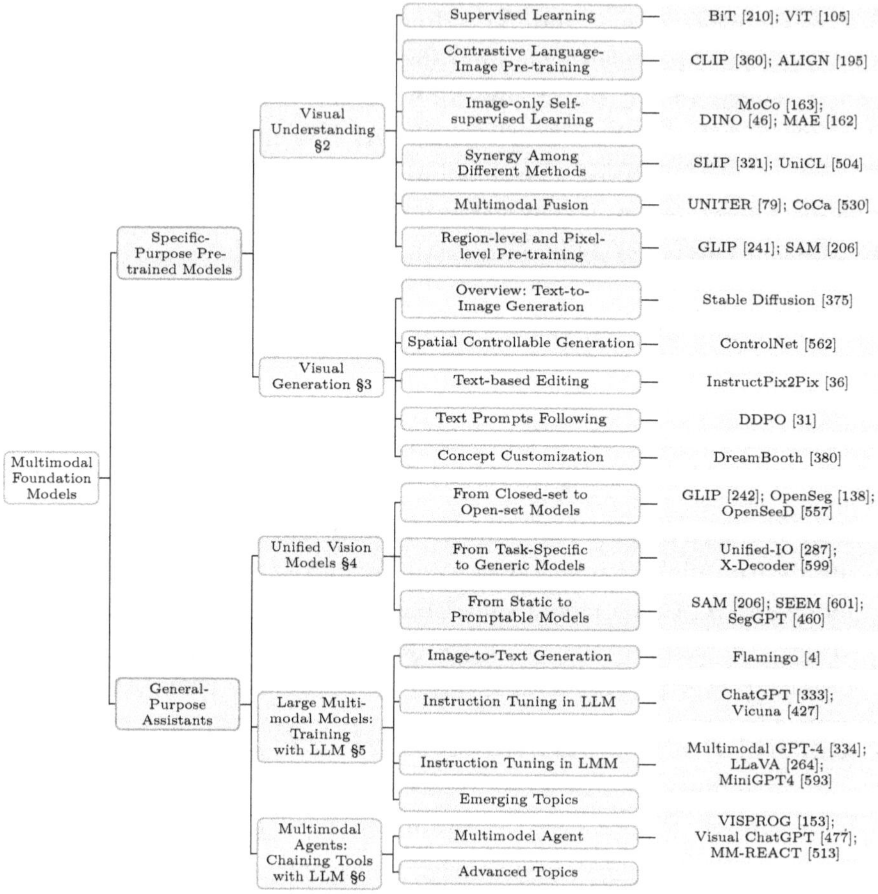

Figure 1.4: An overview of the monograph's structure, detailing Sections 2-6.

multimodal foundation models. The target audiences are graduate students, researchers and professionals who are not experts of multimodal foundation models but are eager to develop perspectives and learn the trends in the field. The structure of this monograph is illustrated in Figure 1.4. It consists of 7 sections.

- Section 1 introduces the landscape of multimodal foundation model research, and presents a historical view on the transition of research from specialists to general-purpose assistants.

- Section 2 introduces different ways to consume visual data, with a focus on how to learn a strong image backbone.
- Section 3 describes how to produce visual data that aligns with human intents.
- Section 4 describes how to design unified vision models, with an interface that is interactive and promptable, especially when LLMs are not employed.
- Section 5 describes how to train an LLM in an end-to-end manner to consume visual input for understanding and reasoning.
- Section 6 describes how to chain multimodal tools with an LLM to enable new capabilities.
- Section 7 concludes the monograph and discusses research trends.

Relations among Sections 2-6. Sections 2-6 are the core sections of this survey. An overview of the structure for these sections are provided in Figure 1.3. We start with a discussion of two typical multimodal foundation models for specific tasks, including visual understanding in Section 2 and visual generation in Section 3. As the notion of multimodal foundation models are originally based on visual backbone/representation learning for understanding tasks, we first present a comprehensive review to the transition of image backbone learning methods, evolving from early supervised methods to the recent language-image contrastive methods, and extend the discussion on image representations from image-level to region-level and pixel-level (Section 2). Recently, generative AI is becoming increasingly popular, where vision generative foundation models have been developed. In Section 3, we discuss large pre-trained text-to-image models, and various ways that the community leverage the generative foundation models to develop new techniques to make them better aligned with human intents. Inspired by the recent advances in NLP that LLMs serve as general-purpose assistants for a wide range of language tasks in daily life, the computer vision community has been anticipating and attempting to build general-purpose visual assistants. We discuss three different ways to build general-purpose assistants. Inspired by the spirit of LLMs,

Section 4 focuses on unifying different vision models of understanding and generation without explicitly incorporating LLMs in modeling. In contrast, Section 5 and Section 6 focus on embracing LLMs to build general-purpose visual assistants, by explicitly augmenting LLMs in modeling. Specifically, Section 5 describes end-to-end training methods, and Section 6 focuses on training-free approaches that chain various vision models to LLMs.

How to read the monograph. Different readers have different backgrounds, and may have different purposes of reading this monograph. Here, we provide some guidance.

- Each section is mostly self-contained. If you have a clear goal and a clear research direction that you want to focus on, then just jump to the corresponding section. For example, if you are interested in building a mini prototype using OpenAI's multimodal GPT-4, then you can directly jump to Section 5.
- If you are a beginner of multimodal foundation models, and are interested in getting a glimpse of the cutting-edge research, we highly recommend that you read the whole monograph section by section in order, as the early sections serve as the building blocks of later sections, and each section provides the description of the key concepts to help you understand the basic ideas, and a comprehensive literature review that to help you grasp the landscape and state of the art.
- If you already have rich experience in multimodal foundation models and are familiar with the literature, feel free to jump to specific sections you want to read. In particular, we include in most sections a section to discuss advanced topics and sometimes provide our own perspectives, based on the up-to-date literature. For example, in Section 6, we discuss several important aspects of multimodal agents in tool use, including tool creation and its connection to retrieval-augmented methods.

1.4 Related Materials: Slide Decks and Pre-recorded Talks

This survey extends what we present in the CVPR 2023 tutorial by covering the most recent advances in the field. Below, we provide a list of slide decks and pre-recorded talks, which are related to the topics in each section, for references.

- **Section 2**: Visual and Vision-Language Pre-training (Youtube, Bilibili)
- **Section 3**: Alignments in Text-to-Image Generation (Youtube, Bilibili)
- **Section 4**: From Representation to Interface: The Evolution of Foundation for Vision Understanding (Youtube, Bilibili)
- **Section 5**: Large Multimodal Models (Youtube, Bilibili)
- **Section 6**: Multimodal Agents: Chaining Multimodal Experts with LLMs (Youtube, Bilibili)

2

Visual Understanding

Over the past decade, the research community has devoted significant efforts to study the acquisition of high-quality, general-purpose image representations. This is essential to build vision foundation models, as pre-training a strong vision backbone to learn image representations is fundamental to all types of computer vision downstream tasks, ranging from image-level (*e.g.*, image classification [212], image-text retrieval [122], image captioning [76]), region-level (*e.g.*, object detection [140], phrase grounding [353]), to pixel-level (*e.g.*, semantic/instance/panoptic segmentation [160], [205], [284]) tasks. Figure 2.1 illustrate an AI that observes and understands the visual world.

In this section, we present how image representations can be learned, either using supervision signals mined inside the images, or through using language supervision of image-text datasets mined from the Web. Specifically, Section 2.1 presents an overview of different learning paradigms, including supervised pre-training, contrastive language-image pre-training (CLIP), and image-only self-supervised learning. Section 2.2 discusses supervised pre-training. Section 2.3 focuses on CLIP. Section 2.4 discusses image-only self-supervised learning, including contrastive learning, non-contrastive learning, and masked image modeling. Given the various

Figure 2.1: Illustration of a visual encoder to observe and understand the world

learning approaches to training vision foundation models, Section 2.5 reviews how they can be incorporated for better performance. Lastly, Section 2.6 discusses how vision foundation models can be used for finer-grained visual understanding tasks, such as fusion-encoder-based pre-training for image captioning and visual question answering that require multimodal fusion, region-level pre-training for grounding, and pixel-level pre-training for segmentation.

2.1 Overview

There is a vast amount of literature on various methods of learning general-purpose vision backbones. As illustrated in Figure 2.2, we group these methods into three categories, depending on the types of supervision signals used to train the models, including:

- **Label supervision**: Arguably, the most well-studied image representation learning methods are based on label supervisions (typically in the form of image classification) [415], where datasets like ImageNet [212] and ImageNet21K [374] have been popular, and larger-scale proprietary datasets are also used in industrial labs [406], [415], [482], [548].

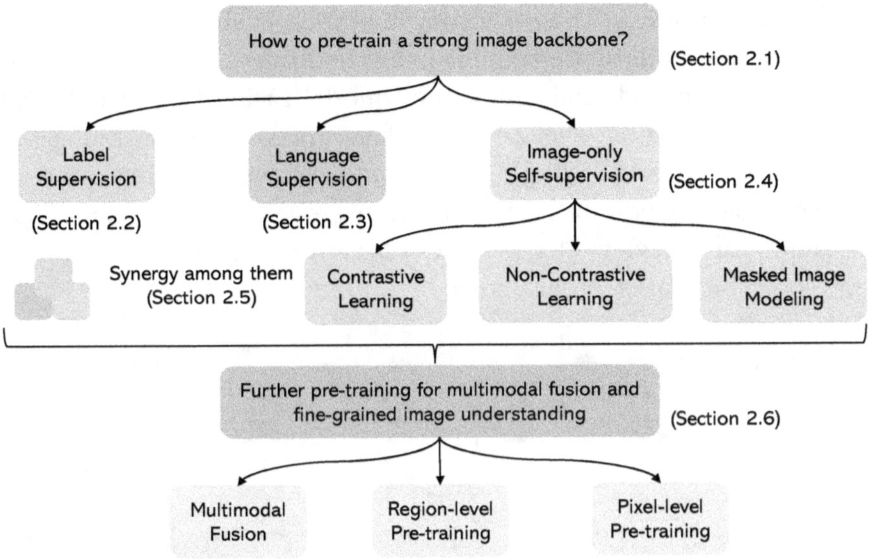

Figure 2.2: An overview of the structure of Section 2.

- **Language supervision**: Another popular approach to learning image representations leverages weakly supervised signals from text, which is easy to acquire in large scale. For instance, CLIP [360] and ALIGN [195] are pre-trained using a contrastive loss and billions of image-text pairs mined from the internet. The resultant models achieve strong zero-shot performance on image classification and image-text retrieval, and the learned image and text encoders have been widely used for various downstream tasks and allow traditional computer vision models to perform open-vocabulary CV tasks [101], [137], [146], [253], [310], [357], [557], [599].

- **Image-only self-supervision**: There is also a vast amount of literature on exploring image-only self-supervised learning methods to learn image representations. As the name indicates, the supervision signals are mined from the images themselves, and popular methods range from contrastive learning [62], [163], non-contrastive learning [46], [77], [144], to masked image modeling [26], [162].

An illustration of these learning methods is shown in Figure 2.3. Besides the methods of pre-training image backbones, we will also discuss pre-training methods that allow multimodal fusion (*e.g.*, CoCa [530], Flamingo [4]), region-level and pixel-level image understanding (*e.g.*, GLIP [241] and SAM [206]). These methods typically rely on a pre-trained image encoder or a pre-trained image-text encoder pair. Figure 2.4 shows an overview of the topics covered in this section and some representative works in each topic.

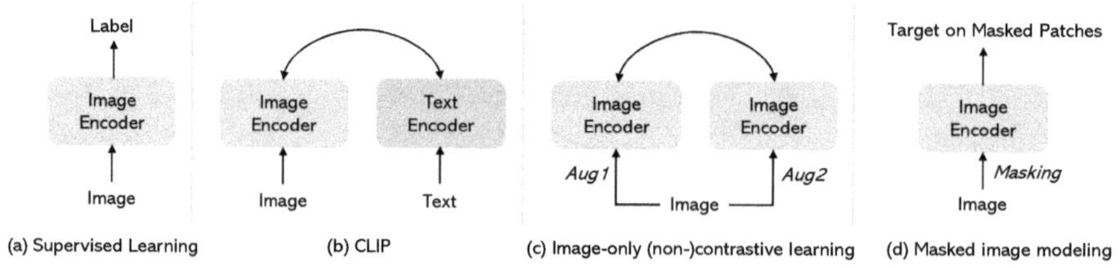

Figure 2.3: A high-level overview of different approaches to learn general image representations, including supervised learning [212], contrastive language-image pre-training [195], [360], and image-only self-supervised learning, including contrastive learning [62], [163], non-contrastive learning [77], [144], and masked image modeling [26], [162].

2.2 Supervised Pre-training

Supervised pre-training on large-scale human-labeled datasets, such as ImageNet [212] and ImageNet21K [374], has emerged as a widely adopted approach to acquiring transferable visual representations. It aims to map an image to a discrete label, which is associated with a visual concept. This approach has greatly expedited progress in designing various vision backbone architectures (*e.g.*, AlexNet [212], ResNet [166], vision transformer [105], and Swin transformer [280]), and is the testbed for all the modern vision backbones. It also powered computer vision tasks across the whole spectrum, ranging from image classification, object detection/segmentation, visual question answering, image captioning, to video action recognition. However, the effectiveness of learned representations is often limited by the scale and diversity of supervisions in pre-training datasets, as human annotation is expensive.

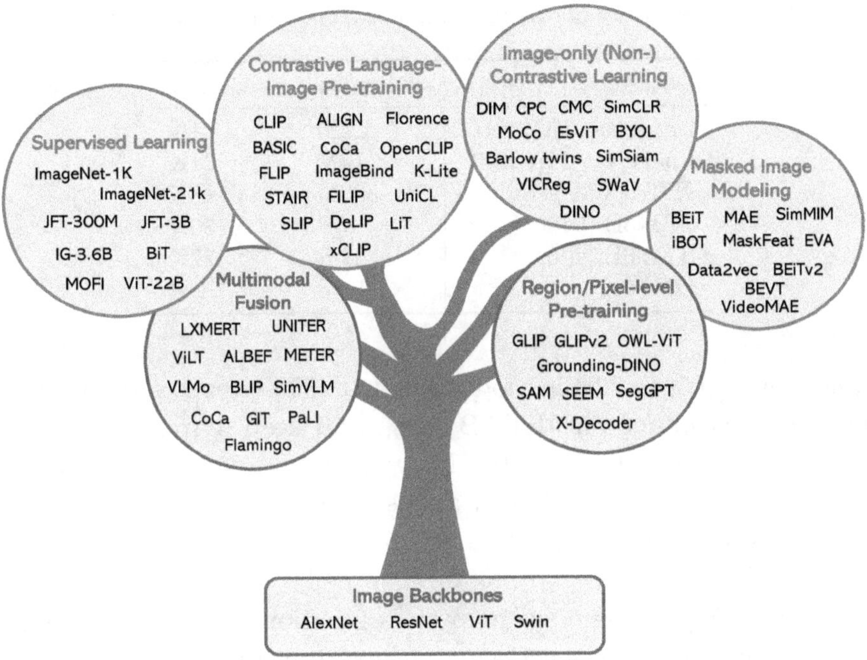

Figure 2.4: An overview of the topics covered in this section and representative works in each topic. We start from supervised learning and CLIP, and then move on to image-only self-supervised learning, including contrastive learning, non-contrastive learning, and masked image modeling. Lastly, we discuss pre-training methods that empower multimodal fusion, region-level and pixel-level image understanding.

Large-scale datasets. For larger-scale pre-training, noisy labels can be derived in large quantities from image-text pairs crawled from the Web. Using noisy labels, many industrial labs have successfully constructed comprehensive classification datasets using semi-automatic pipelines, such as JFT [415], [548] and I2E [482], or by leveraging proprietary data like Instagram hashtags [406]. The statistics of existing large-scale image classification datasets are shown in Table 2.1. The labels are typically in the form of fine-grained image entities with a long-tailed distribution. Though classical, this approach has been very powerful for learning universal image representations. For example, JFT-300M [415] has been used for training the BiT ("Big Transfer") models [210], and JFT-3B [548] has been used to scale up the training of a plain vision

Table 2.1: Statistics of existing large-scale image classification datasets.

Dataset	# Images	# Classes
ImageNet-1K [381]	1.2M	1K
ImageNet-21K [374]	14M	21K
JFT-300M [415]	300M	18K
JFT-3B [548]	3B	30K
IG-3.6B [406]	3.6B	27K
I2E [482]	1.1B	2M

transformer [105] to 22B in model size. LiT [550] proposes to first learn the image backbone on JFT-3B [548], and keep it frozen and learn another text tower to align the image and text embedding space to make the model open-vocabulary and is capable of performing zero-shot image classification.

Model training. There are many loss functions that can be used to promote embedding properties (*e.g.*, separability) [324]. For example, the large margin loss [444] is used for MOFI training [482]. Furthermore, if the datasets have an immense number of labels (can potentially be over 2 million as in MOFI [482]), predicting all the labels in each batch becomes computationally costly. In this case, a fixed number of labels is typically used for each batch, similar to sampled softmax [158].

2.3 Contrastive Language-Image Pre-training

2.3.1 Basics of CLIP Training

Language is a richer form of supervision than classical closed-set labels. Rather than deriving noisy label supervision from web-crawled image-text datasets, the alt-text can be directly used for learning transferable image representations, which is the spirit of contrastive language-image pre-training (CLIP) [360]. In particular, models trained in this way, such as ALIGN [195], Florence [539], BASIC [350], and OpenCLIP [190], have showcased impressive zero-shot image classification and image-text retrieval capabilities by mapping images and text into a shared embedding space. Below, we discuss how the CLIP model is pre-trained and used for zero-shot prediction.

- **Training**: As shown in Figure 2.5 (1), CLIP is trained via simple contrastive learning. CLIP is an outstanding example of "*simple algorithms that scale well*" [249]. To achieve satisfactory performance, model training needs to be scaled along three dimensions: batch size, data size, and model size [350]. Specifically, the typical batch size used for CLIP training can be 16k or 32k. The number of image-text pairs in the pre-training datasets is frequently measured in billions rather than millions. A vision transformer trained in this fashion can typically vary from 300M (Large) to 1B (giant) in model size.

- **Zero-shot prediction**: As shown in Figure 2.5 (2) and (3), CLIP empowers zero-shot image classification via reformatting it as a retrieval task and considering the semantics behind labels. It can also be used for zero-shot image-text retrieval by its design. Besides this, the aligned image-text embedding space makes it possible to make all the traditional vision models open vocabulary and has inspired a rich line of work on open-vocabulary object detection and segmentation [241], [557], [559], [599].

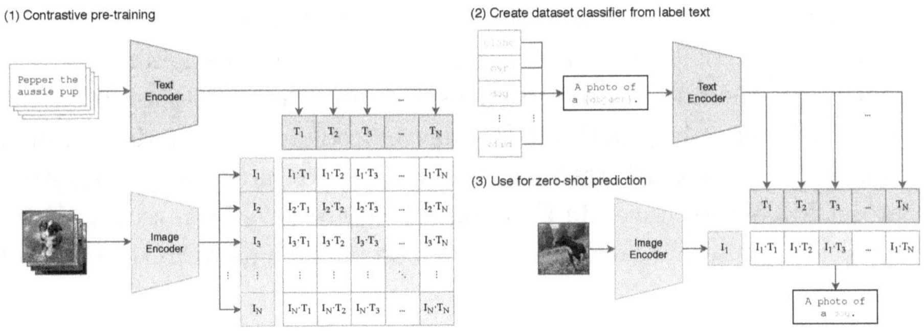

Figure 2.5: Illustration of contrastive language-image pre-training, and how the learned model can be used for zero-shot image classification. Image credit: [360].

2.3.2 CLIP Variants

Since the birth of CLIP, there have been tons of follow-up works to improve CLIP models, as to be discussed below. We do not aim to provide a comprehensive literature review of all the methods, but focus on a selected set of topics.

Data scaling up. Data is the fuel for CLIP training. For example, OpenAI's CLIP was trained on 400M image-text pairs mined from the web, while ALIGN used a proprietary dataset consisting of 1.8B image-text pairs. In BASIC [350], the authors have carefully studied the scaling among three dimensions: batch size, data size, and model size. However, most of these large-scale datasets are not publicly available, and training such models requires massive computing resources.

In academic settings, researchers [227] have advocated the use of a few millions of image-text pairs for model pre-training, such as CC3M [392], CC12M [50], YFCC [428]. Relatively small-scale image-text datasets that are publicly available include SBU [337], RedCaps [96], and WIT [411]. Large-scale public available image-text datasets include Shutterstock [327], LAION-400M [386], COYO-700M [38], and LAION-2B [385], to name a few. For example, LAION-2B [385] has been used by researchers to study the reproducible scaling laws for CLIP training [82].

Interestingly, in search of the next-generation image-text datasets, in DataComp [125], instead of fixing the dataset and designing different algorithms, the authors propose to select and rank datasets using the fixed CLIP training method. Besides paired image-text data mined from the Web for CLIP training, inspired by the interleaved image-text dataset M3W introduced in Flamingo [4], there have been recent efforts of collecting interleaved image-text datasets, such as MMC4 [596] and OBELISC [219].

Model design and training methods. CLIP training has been significantly improved. Below, we review some representative works.

- **Image tower**: On the image encoder side, FLIP [249] proposes to scale CLIP training via masking. By randomly masking out image patches with a high masking ratio, and only encoding the visible patches as in MAE [162], the authors demonstrate that masking can improve training efficiency without hurting the performance. The method can be adopted for all CLIP training. Cao *et al.* [42] found that filtering out samples that contain text regions in the image improves CLIP training efficiency and robustness.

- **Language tower**: On the language encoder side, K-Lite [394] proposes to use external knowledge in the form of Wiki definition of entities together with the original alt-text for contrastive pre-training. Empirically, the use of enriched text descriptions improves the CLIP performance. LaCLIP [114] shows that CLIP can be improved via rewriting the noisy and short alt-text using large language models such as ChatGPT.

- **Interpretability**: The image representation is typically a dense feature vector. In order to improve the interpretability of the shared image-text embedding space, STAIR [52] proposes to map images and text to a high-dimensional, sparse, embedding space, where each dimension in the sparse embedding is a (sub-)word in a large dictionary in which the predicted non-negative scalar corresponds to the weight associated with the token. The authors show that STAIR achieves better performance than the vanilla CLIP with improved interpretability.

- **More modalities**: The idea of contrastive learning is general, and can go beyond just image and text modalities. For example, as shown in Figure 2.6, ImageBind [139] proposes to encode six modalities into a common embedding space, including images, text, audio, depth, thermal, and IMU modalities. In practice, a pre-trained CLIP model is used and kept frozen during training, which indicates that other modality encoders are learned to align to the CLIP embedding space, so that the trained model can be applied to new applications such as audio-to-image generation and multimodal LLMs (*e.g.*, PandaGPT [414]).

Objective function. The use of contrastive loss alone is powerful, especially when the model is scaled up. However, other objective functions can also be applied.

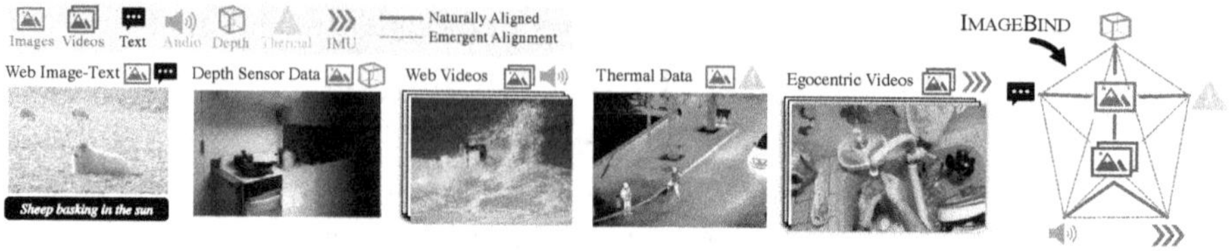

Figure 2.6: ImageBind [139] proposes to link a total of six modalities into a common embedding space via leveraging pre-trained CLIP models, enabling new emergent alignments and capabilities. Image credit: [139].

- **Fine-grained supervision**: Instead of using a simple dot-product to calculate the similarity of an image-text pair, the supervision can be made more fine-grained via learning word-patch alignment. In FILIP [516], the authors propose to first compute the loss by calculating the token-wise similarity, and then aggregating the matrix by max-pooling for word-patch alignment.

- **Contrastive captioner**: Besides the contrastive learning branch, CoCa [530] (shown in Figure 2.7(c)) adds a generative loss to improve performance and allow new capabilities that require multimodal fusion (*e.g.*, image captioning and VQA). This is similar to many fusion-encoder-based vision-language models such as ALBEF [236], but with the key difference in that CoCa aims to learn a better image encoder from scratch. A detailed discussion on multimodal fusion is in Section 2.6.1.

- **Captioning loss alone**: How about using the captioning loss alone to pre-train an image encoder? Actually, before CLIP was invented, VirTex [95] (shown in Figure 2.7(b)) and ICMLM [383] learn encoders using a single image captioning loss, but the scale is very small (restricted to COCO images) and the performance is poor. CLIP also shows that contrastive pre-training is a much better choice. In SimVLM [468], the authors found that the learned image encoder was not as competitive as CLIP. However, in the recent work Cap/CapPa [434], the authors argue that image captioners are scalable vision learners, too. Captioning can exhibit the same or even better scaling behaviors.

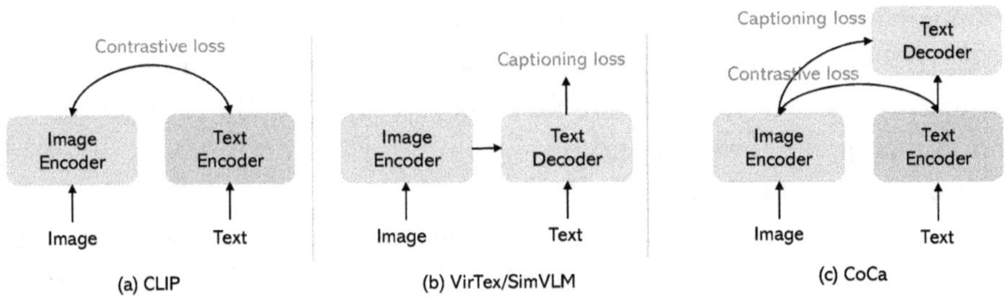

Figure 2.7: A high-level comparison of contrastive loss and captioning loss for image encoder pre-training. (a) CLIP [360] uses contrastive loss alone for pre-training, which enables zero-shot image classification and has demonstrated strong scaling behavior. (b) VirTex [95] uses captioning loss alone for pre-training. SimVLM [468] uses prefix language modeling for pre-training in a much larger scale. The model architecture is similar to multimodal language models (*e.g.*, GIT [446] and Flamingo [4]), but VirTex and SimVLM aim to pre-train the image encoder from scratch. (c) CoCa [530] uses both contrastive and captioning losses for pre-training. The model architecture is similar to ALBEF [236], but CoCa aims to pre-train the image encoder from scratch, instead of using a pre-trained one.

- **Sigmoid loss for language-image pre-training**: Unlike standard contrastive learning with softmax normalization, [549] uses a simple pairwise sigmoid loss for image-text pre-training, which operates on image-text pairs and does not require a global view of the pairwise similarities for normalization. The authors show that the use of simple sigmoid loss can also achieve strong performance on zero-shot image classification.

2.4 Image-Only Self-Supervised Learning

Now, we shift our focus to image-only self-supervised learning, and divide the discussion into three parts: (*i*) contrastive learning, (*ii*) non-contrastive learning, and (*iii*) masked image modeling.

2.4.1 Contrastive and Non-contrastive Learning

Contrastive learning. The core idea of contrastive learning [9], [158] is to promote the positive sample pairs and repulse the negative sample pairs. Besides being used in CLIP, contrastive learning has also been a popular concept in self-supervised image representation learning [62], [75], [163], [311], [429], [484], [521]. It has been shown that the contrastive objective, known as the InfoNCE loss [331], can be interpreted as maximizing the lower bound of mutual information between different views of the data [17], [169], [171].

In a nutshell, all the image-only contrastive learning methods (*e.g.*, SimCLR [62], see Figure 2.8(a), MoCo [163], SimCLR-v2 [63], MoCo-v2 [75]) share the same high-level framework, detailed below.

- Given one image, two separate data augmentations are applied;
- A base encoder is followed by a project head, which is trained to maximize agreement using a contrastive loss (*i.e.*, they are from the same image or not);
- The project head is thrown away for downstream tasks.

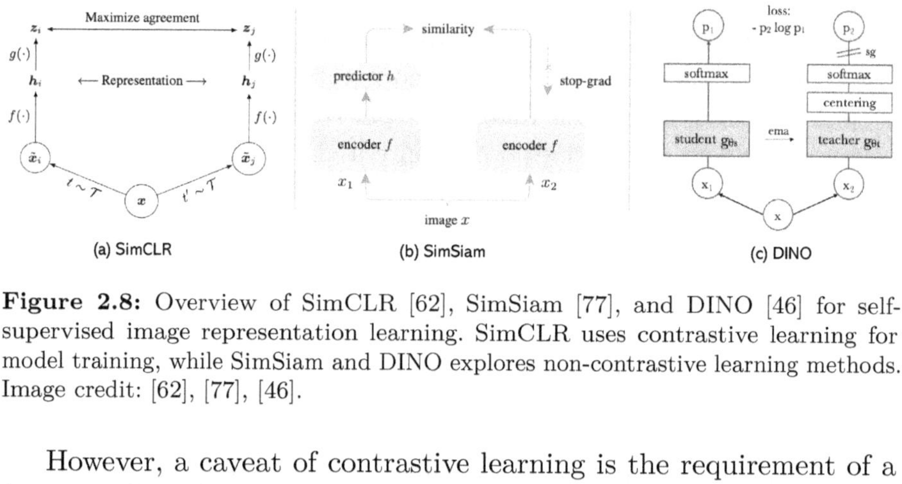

Figure 2.8: Overview of SimCLR [62], SimSiam [77], and DINO [46] for self-supervised image representation learning. SimCLR uses contrastive learning for model training, while SimSiam and DINO explores non-contrastive learning methods. Image credit: [62], [77], [46].

However, a caveat of contrastive learning is the requirement of a large number of negative samples. These samples can be maintained in a memory bank [484], or directly from the current batch [62], which suggests the requirement of a large batch size. MoCo [163] maintains a queue of negative samples and turns one branch into a momentum encoder to improve the consistency of the queue. Initially, contrastive learning was primarily studied for pre-training convolutional networks.

However, with the rising popularity of vision transformers (ViT), researchers have also explored its application in the context of ViT. [78], [229], [489].

Non-contrastive learning. Recent self-supervised learning methods do not depend on negative samples. The use of negatives is replaced by asymmetric architectures (*e.g.*, BYOL [144], SimSiam [77]), dimension de-correlation (*e.g.*, Barlow twins [544], VICReg [28], Whitening [111]), and clustering (*e.g.*, SWaV [45], DINO [46], [6], [10], [44], [443]), *etc.*

For example, as illustrated in Figure 2.8(b), in SimSiam [77], two augmented views of a single image are processed by an identical encoder network. Subsequently, a prediction MLP is applied to one view, while a stop-gradient operation is employed on the other. The primary objective of this model is to maximize the similarity between the two views. It is noteworthy that SimSiam relies on neither negative pairs nor a momentum encoder.

Another noteworthy method, known as DINO [46] and illustrated in Figure 2.8(c), takes a distinct approach. DINO involves feeding two distinct random transformations of an input image into both the student and teacher networks. Both networks share the same architecture but have different parameters. The output of the teacher network is centered by computing the mean over the batch. Each network outputs a feature vector that is normalized with a temperature softmax applied to the feature dimension. The similarity between these features is quantified using a cross-entropy loss. Additionally, a stop-gradient operator is applied to the teacher network to ensure that gradients propagate exclusively through the student network. Moreover, DINO updates the teacher's parameters using an exponential moving average of the student's parameters.

2.4.2 Masked Image Modeling

Masked language modeling [98] is a powerful pre-training task that has revolutionized the NLP research. To mimic the success of BERT pre-training for NLP, the pioneering work BEiT [26], as illustrated in Figure 2.9, proposes to perform masked image modeling (MIM) to pre-train image transformers. Specifically,

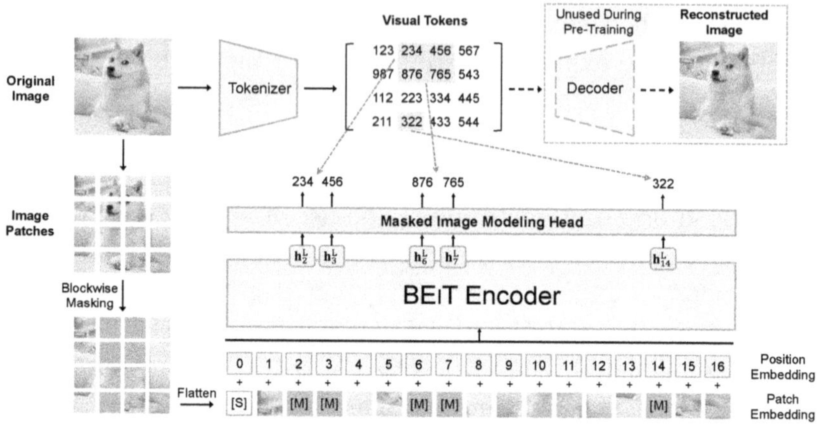

Figure 2.9: Overview of BEiT pre-training for image transformers. Image credit: [26].

- **Image tokenizer**: In order to perform masked token prediction, an image tokenizer is required to tokenize an image into discrete visual tokens, so that these tokens can be treated just like an additional set of language tokens. Some well-known learning methods for image tokenziers include VQ-VAE [330], VQ-VAE-2 [370], VQ-GAN [112], ViT-VQGAN [529], *etc.* These image tokenizers have also been widely used for autoregressive image generation, such as DALLE [367], Make-A-Scene [126], Parti [531], to name a few.

- **Mask-then-predict**: The idea of MIM is conceptually simple: models accept the corrupted input image (*e.g.*, via random masking of image patches), and then predict the target of the masked content (*e.g.*, discrete visual tokens in BEiT). As discussed in iBOT [588], this training procedure can be understood as knowledge distillation between the image tokenizer (which serves as the teacher) and the BEiT encoder (which serves as the student), while the student only sees partial of the image.

Targets. In [346], the authors have provided a unified view of MIM: a teacher model, a normalization layer, a student model, an MIM head, and a proper loss function. The most significant difference among all these models lies in the reconstruction targets, which can be pixels, discrete image tokens, features from pre-trained models, and outputs from the momentum updated teacher. Specifically, the targets can be roughly grouped into two categories.

- **Low-level pixels/features as targets**: MAE [162], SimMIM [490], ConvMAE [132], HiViT [571], and GreenMIM [181] leverage either original or normalized pixel values as the target for MIM. These methods have typically explored the use of a plain Vision Transformer [105] or the Swin Transformer [280] as the backbone architecture. MaskFeat [470] introduced the Histogram of Oriented Gradients (HOG) feature descriptor as the target for MIM (see Figure 2.10(b)). Meanwhile, Ge^2-AE [262] employed both pixel values and frequency information obtained from the 2D discrete Fourier transform as the target. Taking MAE [162] as an example (Figure 2.10(a)), the authors show that using pixel values as targets works particularly well. Specifically, a large random subset of images (*e.g.*, 75%) is masked out; then, the image encoder is only applied to visible patches, while mask tokens are introduced after the encoder. It was shown that such pre-training is especially effective for object detection and segmentation tasks, which require finer-grained image understanding.
- **High-level features as targets**: BEiT [26], CAE [74], SplitMask [329], and PeCo [104] involve the prediction of discrete tokens using learned image tokenizers. MaskFeat [470] takes a different approach by proposing direct regression of high-level features extracted from models like DINO [46] and DeiT [432]. Expanding this idea, MVP [473] and EVA [116] make feature prediction using image features from CLIP as target features. Additionally, other methods such as data2vec [18], MSN [10], ConMIM [523], SIM [424], and BootMAE [103] propose to construct regression feature targets by leveraging momentum-updated teacher models to enhance online learning. The choice of loss functions depends on the nature of the targets: cross-entropy loss is typically used when the targets are discrete tokens, while ℓ_1, ℓ_2, or cosine similarity losses are common choices for pixel values or continuous-valued features.

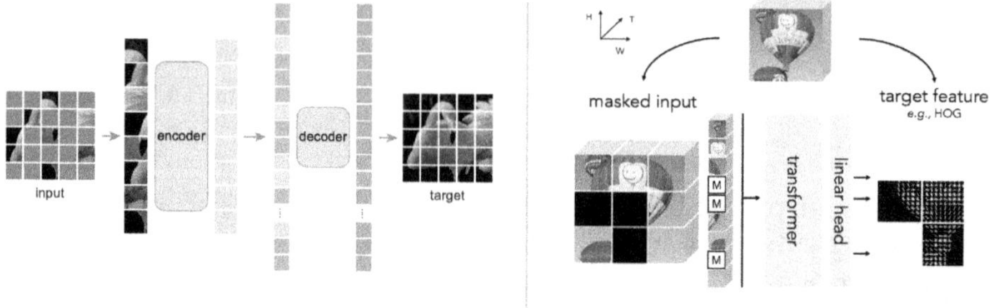

(a) MAE (b) MaskFeat

Figure 2.10: Illustration of Masked Autoencoder (MAE) [162] that uses raw pixel values for MIM training, and MaskFeat [470] that uses different features as the targets. HOG, a hand-crafted feature descriptor, was found to work particularly well in terms of both performance and efficiency. Image credit: [162] and [470].

MIM for video pre-training. Naturally, there are recent works on extending MIM to video pre-training. Prominent examples include BEVT [452], MAE as spatiotemporal learner [118], VideoMAE [431], and VideoMAEv2 [450]. Taking [118] as an example, this paper studies a conceptually simple extension of MAE to video pre-training via randomly masking out space-time patches in videos and learns an autoencoder to reconstruct them in pixels. Interestingly, the authors found that MAE learns strong video representations with almost no inductive bias on space-time, and spacetime-agnostic random masking performs the best, with an optimal masking ratio as high as 90%.

Lack of learning global image representations. MIM is an effective pre-training method that provides a good parameter initialization for further model finetuning. However, the vanilla MIM pre-trained model does not learn a global image representation. In iBOT [588], the authors propose to enhance BEiT [26] with a DINO-like self-distillation loss [46] to force the `[CLS]` token to learn global image representations. The same idea has been extended to DINOv2 [336].

Scaling properties of MIM. MIM is scalable in terms of model size. For example, we can perform MIM pre-training of a vision transformer with billions of parameters. However, the scaling property with regard to data size is less clear. There are some recent works that aim to understand the data scaling of MIM [285], [491]; however, the data scale is limited to millions of images, rather than billions, except [405] that studies the effectiveness of MAE as a so-called "pre-pretraining" method for billion-scale data. Generally, MIM can be considered an effective regularization method that helps initialize a billion-scale vision transformer for downstream tasks; however, whether or not scaling the MIM pre-training to billion-scale image-only data requires further exploration.

2.5 Synergy Among Different Learning Approaches

Till now, we have reviewed different approaches to pre-training image backbones, especially for vision transformers. Below, we use CLIP as the anchor point, and discuss how CLIP can be combined with other learning methods.

Combining CLIP with label supervision. Noisy labels and text supervision can be jointly used for image backbone pre-training. Some representative works are discussed below.

- UniCL [503] proposes a principled way to use image-label and image-text data together in a joint image-text-label space for unified contrastive learning, and Florence [539] is a scaled-up version of UniCL. See Figure 2.11 for an illustration of the framework.
- LiT [550] uses a pre-trained ViT-g/14 image encoder learned from supervised pre-training on the JFT-3B dataset, and then makes the image encoder open-vocabulary by learning an additional text tower via contrastive pre-training on image-text data. Essentially, LiT teaches a text model to read out good representations from a pre-trained image model for new tasks.
- MOFI [482] proposes to learn image representations from 1 billion noisy entity-annotated images, and uses both image classification and contrastive losses for model training. For image classification, entities associated with each image are considered as labels, and

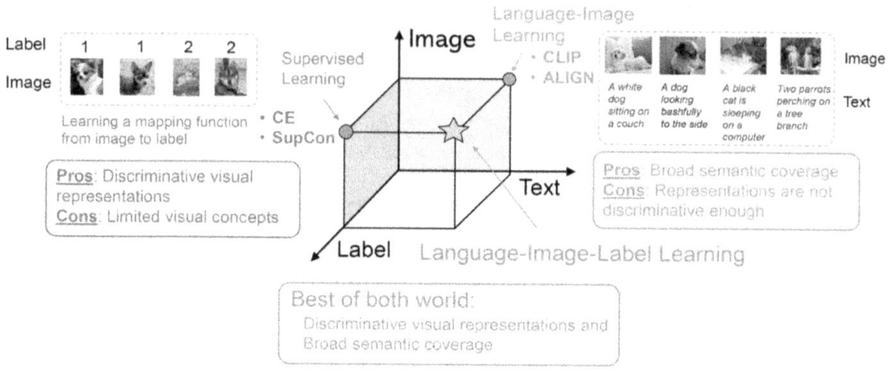

Figure 2.11: Overview of UniCL [503] that performs unified contrastive pre-training on image-text and image-label data. Image credit: [503].

supervised pre-training on a large number of entities is conducted; for constrastive pre-training, entity names are treated as free-form text, and are further enriched with entity descriptions.

Combining CLIP with image-only (non-)contrastive learning. CLIP can also be enhanced with image-only self-supervision. Specifically,

- SLIP [321] proposes a conceptually simple idea to combine SimCLR [62] and CLIP for model training, and shows that SLIP outperforms CLIP on both zero-shot transfer and linear probe settings. DeCLIP [248] mines self-supervised learning signals on each modality to make CLIP training data-efficient. In terms of image supervision, the SimSam framework [77] is used.

- xCLIP [587] makes CLIP non-contrastive via introducing additional sharpness and smoothness regularization terms borrowed from the image-only non-contrastive learning literature. However, the authors show that only non-contrastive pre-training (nCLIP) is not sufficient to achieve strong performance on zero-shot image classification, and it needs to be combined with the original CLIP for enhanced performance.

Combining CLIP with MIM. There are recent works that aim to combine CLIP and MIM for model training. We group them into two categories.

- **Shallow interaction.** It turns out that image features extracted from CLIP are a good target for MIM training, as the CLIP image features potentially capture the semantics that are missing in MIM training. Along this line of work, as shown in Figure 2.12, MVP [473] proposes to regress CLIP features directly, while BEiTv2 [347] first compresses the information inside CLIP features into discrete visual tokens, and then performs regular BEiT training. Similar use of CLIP features as MIM training target has also been investigated in EVA [116], CAEv2 [572], and MaskDistill [346]. In EVA-02 [116], the authors advocate alternative learning of MIM and CLIP representations. Specifically, an off-the-shelf CLIP model is used to provide a feature target for MIM training; while the MIM pre-trained image backbone is used to initialize CLIP training. The MIM representations are used to finetune various downstream tasks while the learned frozen CLIP embedding enables zero-shot image classification and other applications.
- **Deeper integration.** However, instead of using CLIP as targets for MIM training, if one aims to combine CLIP and MIM for joint model training, MIM does not seem to improve a CLIP model at scale [249], [469].
- Although the combination of CLIP and MIM does not lead to a promising result at the current stage, the combination of BERT and BEiT is very promising, as evidenced in BEiT-3 [457] (see Figure 2.13), where the authors show that masked data modeling can be performed on both image/text and joint image-text data via the design of a multiway transformer, and state-of-the-art performance can be achieved on a wide range of vision and vision-language tasks.

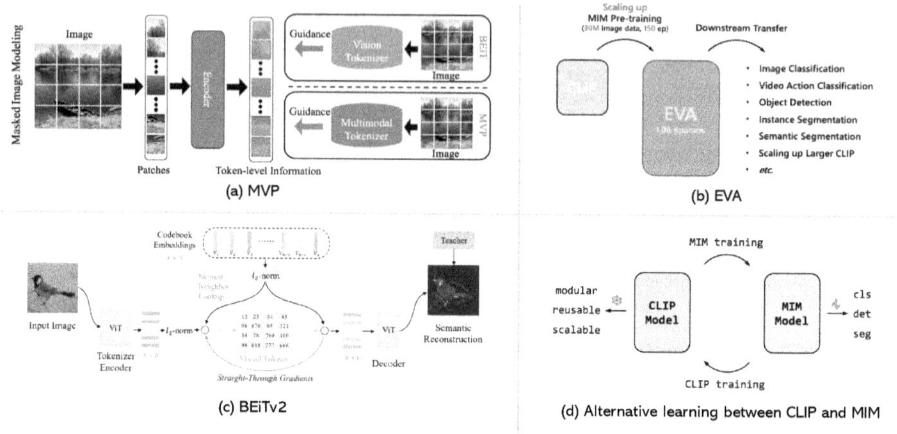

Figure 2.12: Illustration of MVP [473], EVA [116] and BEiTv2 [347]. (a) & (b) MVP and EVA directly regress CLIP features for MIM pre-training. (c) BEiTv2 compresses the information inside CLIP features into discrete visual tokens, and then performing regular BEiT training. (d) Alternative learning between CLIP and MIM. Image credit: [473], [116], [347], [116].

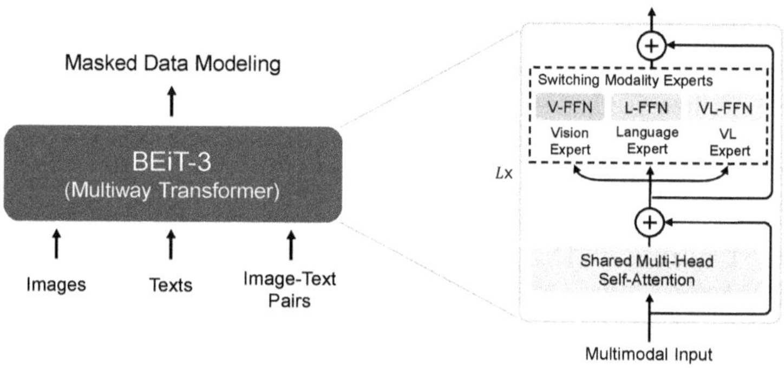

Figure 2.13: Overview of BEiT-3 that performs masked data modeling on both image/text and joint image-text data via a multiway transformer. Image credit: [457].

2.6 Multimodal Fusion, Region-Level and Pixel-Level Pre-training

Till now, we have focused on the methods of pre-training image backbones from scratch, but not on pre-training methods that power multimodal fusion, region-level and pixel-level image understanding. These methods typically use a pre-trained image encoder at the first hand to perform a second-stage pre-training. Below, we briefly discuss these topics.

2.6.1 From Multimodal Fusion to Multimodal LLM

For dual encoders such as CLIP [360], image and text are encoded separately, and modality interaction is only handled via a simple dot product of image and text feature vectors. This can be very effective for zero-shot image classification and image-text retrieval. However, due to the lack of deep multimodal fusion, CLIP alone performs poorly on the image captioning [438] and visual question answering [8] tasks. This requires the pre-training of a fusion encoder, where additional transformer layers are typically employed to model the deep interaction between image and text representations. Below, we review how these fusion-encoder pre-training methods are developed over time.

OD-based models. Most early methods use pre-trained object detectors (ODs) to extract visual features. Among them, ViLBERT [286] and LXMERT [421] use co-attention for multimodal fusion, while methods like VisualBERT [240], Unicoder-VL [232], VL-BERT [413], UNITER [79], OSCAR [246], VILLA [128] and VinVL [563] treat image features as soft prompts of the text input to be sent into a multimodal transformer.

End-to-end models. Now, end-to-end pre-training methods become the mainstream. Some early methods use CNNs to extract image features, such as PixelBERT [188], SOHO [187], and CLIP-ViL [395], while ViLT [203] and ViTCAP [117] directly feed image patch features and text token embeddings into a multimodal transformer. Due to the popularity of vision transformer (ViT), now most methods simply use ViT as the image encoder (*e.g.*, plain ViT [105] and Swin transformer [280]). Prominent examples include ALBEF [236], METER [107], VLMo [458], X-VLM [546], BLIP [235], SimVLM [468], FLAVA [404] and CoCa [530].

An illustration of UNITER [79] and CoCa [530] is shown in Figure 2.14. They serve as two examples of a classical model and a modern model, respectively, which performs pre-training on multimodal fusion. CoCa also performs image backbone pre-training directly, as all the model components are trained from scratch. Please refer to Chapter 3 of [129] for a comprehensive literature review.

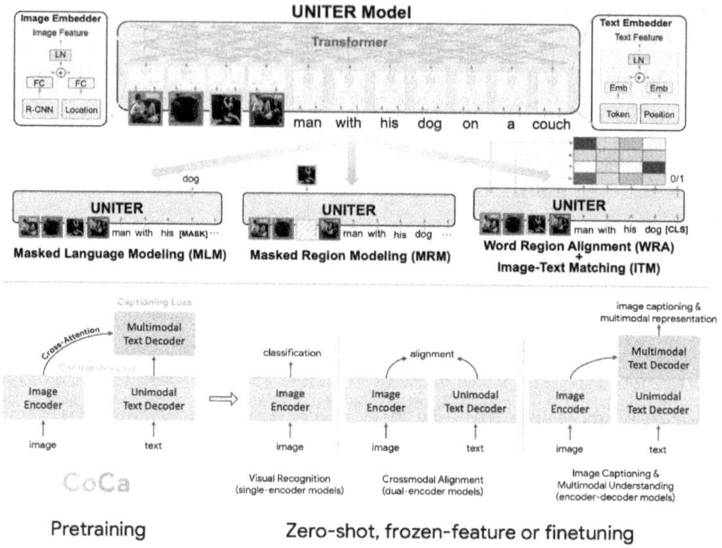

Figure 2.14: Illustration of UNITER [79] and CoCa [530], which serve as a classical and a modern model that performs pre-training on multimodal fusion. CoCa also pre-trains the image backbone from scratch. Specifically, UNITER extracts image features via an off-the-shelf object detector and treat image features as soft prompts of the text input to be sent into a multimodal transformer. The model is pre-trained over a few millions of image-text pairs. For CoCa, an image encoder and a text encoder is used, with a multimodal transformer stacked on top. Both contrastive loss and captioning loss are used for model training, and the model is trained over billions of image-text pairs and JFT data. Image credit: [79], [530].

Trend to multimodal LLM. Instead of using masked language modeling, image-text matching and image-text contrastive learning, SimVLM [468] uses a simple PrefixLM loss for pre-training. Since then, multimodal language models have become popular. Early models focus on large-scale pre-training, such as Flamingo [4], GIT [446], PaLI [71], PaLI-X [70], while recent works focus on using pre-trained LLMs for instruction tuning, such as LLaVA [264] and MiniGPT-4 [593]. A detailed discussion on this topic is provided in Section 5.

2.6.2 Region-Level Pre-training

CLIP learns global image representations via contrastive pre-training. However, for tasks that require fine-grained image understanding such as object detection, CLIP is not enough. Object detection contains two sub-tasks: localization and recognition. (*i*) Localization aims to locate the presence of objects in an image and indicate the position with a bounding box, while (*ii*) recognition determines what object categories are present in the bounding box. By following the reformulation that converts image classification to image retrieval used in CLIP, generic open-set object detection can be achieved.

Specifically, ViLD [146] and RegionCLIP [581] distill knowledge from CLIP with a two-stage detector for zero-shot object detection. In MDETR [198] and GLIP [241] (as shown in Figure 2.15), the authors propose to reformulate detection as a phrase grounding problem, and perform grounded language-image pre-training. GLIPv2 [559] and FIBER [106] further perform unified pre-training for both grounding and vision-language understanding tasks. OVR-CNN [543] finetunes an image-text model to detection on a limited vocabulary and relies on image-text pre-training for generalization to an open vocabulary setting. Detic [590] improves long-tail detection performance with weak supervision by training only the classification head on the examples where only image-level annotations are available. Other works include OV-DETR [542], X-DETR [41], FindIT [214], PromptDet [119], OWL-ViT [310], GRiT [478], to name a few. Recently, Grounding DINO [271] is proposed to marry DINO [556] with grounded pre-training for open-set object detection. Please refer to Section 4.2 for a detailed review of this topic.

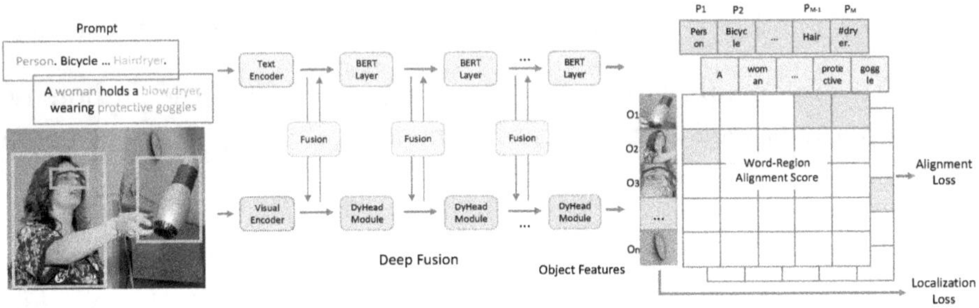

Figure 2.15: Overview of GLIP that performs grounded language-image pre-training for open-set object detection. Image credit: [242].

2.6.3 Pixel-Level Pre-training

The Segment Anything Model (SAM) [206] is a recent vision foundation model for image segmentation that aims to perform pixel-level pre-training. Since its birth, it has attracted wide attention and spurred tons of follow-up works and applications. Below, we briefly review SAM, as a representative work for pixel-level visual pre-training.

As depicted in Figure 2.16, the objective of the Segment Anything project is to develop a foundational vision model for segmentation. This model is designed to be readily adaptable to a wide range of both existing and novel segmentation tasks, such as edge detection, object proposal generation, instance segmentation, open-vocabulary segmentation, and more. This adaptability is seamlessly accomplished through a highly efficient and user-friendly approach, facilitated by the integration of three interconnected components. Specifically,

- **Task.** The authors propose the promptable segmentation task, where the goal is to return a valid segmentation mask given any segmentation prompt, such as a set of points, a rough box or mask, or free-form text.

- **Model.** The architecture of SAM is conceptually simple. It is composed of three main components: (*i*) a powerful image encoder (MAE [162] pre-trained ViT); (*ii*) a prompt encoder (for sparse input such as points, boxes, and free-form text, the CLIP text encoder is used; for dense input such as masks, a convolution operator is used); and (*iii*) a lightweight mask decoder based on transformer.

- **Data.** To acquire large-scale data for pre-training, the authors develop a *data engine* that performs model-in-the-loop dataset annotation.

Concurrent to SAM. Parallel to SAM, many efforts have been made to develop general-purpose segmentation models as well. For example, OneFormer [192] develops a universal image segmentation framework; SegGPT [460] proposes a generalist in-context learning framework that

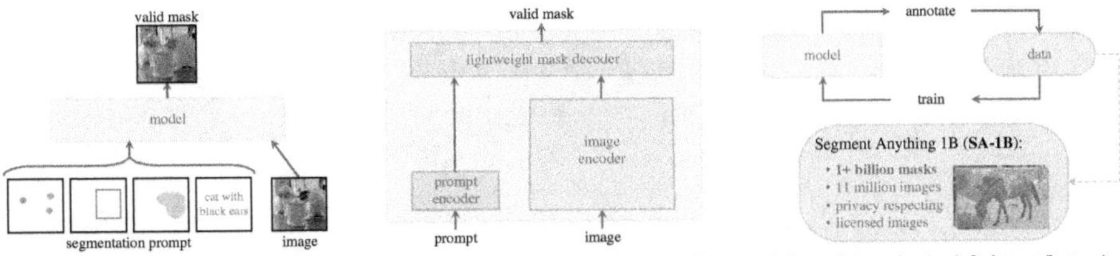

(a) **Task**: promptable segmentation (b) **Model**: Segment Anything Model (**SAM**) (c) **Data**: data engine (top) & dataset (bottom)

Figure 2.16: Overview of the Segment Anything project, which aims to build a vision foundation model for segmentation by introducing three interconnected components: a promptable segmentation task, a segmentation model, and a data engine. Image credit: [206].

unifies different segmentation data formats; SEEM [601] further expands the types of supported prompts that a single segmentation model can handle, including points, boxes, scribbles, masks, texts, and referred regions of another image.

Extensions of SAM. SAM has spurred tons of follow-up works that extend SAM to a wide range of applications, *e.g.*, Inpaint Anything [536], Edit Everything [487], Any-to-Any Style Transfer [272], Caption Anything [454], Track Anything [505], Recognize Anything [230], [575], Count Anything [300], 3D reconstruction [393], medical image analysis [299], [399], [574], [589], *etc.* Additionally, recent works have attempted to develop models for detecting and segmenting anything in the open-vocabulary scenarios, such as Grounding DINO [271] and Grounding-SAM.[1] For a comprehensive review, please refer to [554] and some GitHub repos.[2]

[1]https://github.com/IDEA-Research/Grounded-Segment-Anything
[2]https://github.com/Hedlen/awesome-segment-anything

3

Visual Generation

Visual generation aims to generate high-fidelity visual content, including images, videos, neural radiance fields, 3D point clouds, etc.. This topic is at the core of recently popular artificial intelligence generated content (AIGC), and this ability is crucial in supporting creative applications such as design, arts, and multimodal content creation. It is also instrumental in synthesizing training data to help understand models, leading to the closed loop of multimodal content understanding and generation. To make use of visual generation, it is critical to produce visual data that is strictly aligned with human intents. These intentions are fed into the generation model as input conditions, such as class labels, texts, bounding boxes, layout masks, among others. Given the flexibility offered by open-ended text descriptions, text conditions (including text-to-image/video/3D) have emerged as a pivotal theme in conditional visual generation. Figure 3.1 illustrates a visual generator AI to synthesize the visual world.

In this section, we describe how to align with human intents in visual generation, with a focus on image generation. We start with the overview of the current state of text-to-image (T2I) generation in Section 3.1, highlighting its limitations concerning alignment with human intents.

Figure 3.1: Illustration of a visual generator AI to synthesize the visual world.

The core of this section is dedicated to reviewing the literature on four targeted areas that aim at enhancing alignments in T2I generation, *i.e.*, spatial controllable T2I generation in Section 3.2, text-based image editing in Section 3.3, better following text prompts in Section 3.4, and concept customization in T2I generation in Section 3.5. At the end of each subsection, we share our observations on the current research trends and short-term future research directions. These discussions coalesce in Section 3.6, where we conclude the section by considering future trends. Specifically, we envision the development of a generalist T2I generation model, which can better follow human intents, to unify and replace the four separate categories of alignment works.

3.1 Overview

3.1.1 Human Alignments in Visual Generation

AI Alignment research in the context of T2I generation is the field of study dedicated to developing image generation models that can easily follow human intents to synthesize the desired generated visual content. Current literature typically focuses on one particular weakness of vanilla T2I models that prevents them from accurately producing images that align with human intents. This section delves into four commonly studied issues, as summarized in Figure 3.2 (a) and follows.

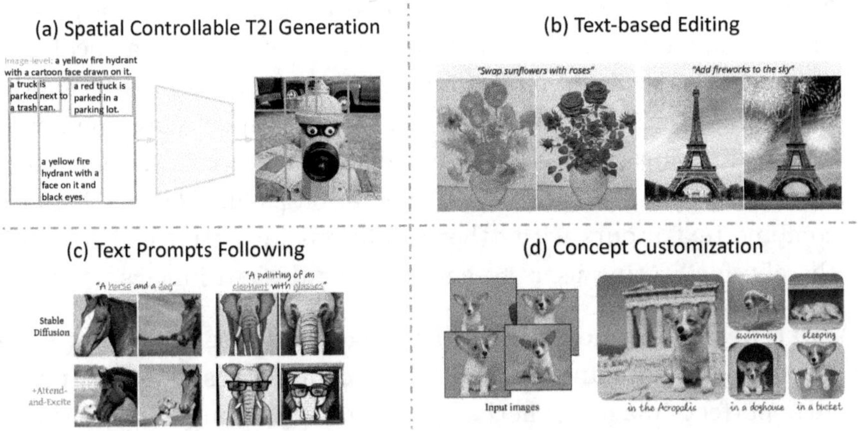

(a) An overview of topics on human alignment for generative foundation models. Image credit: [36], [51], [380], [511].

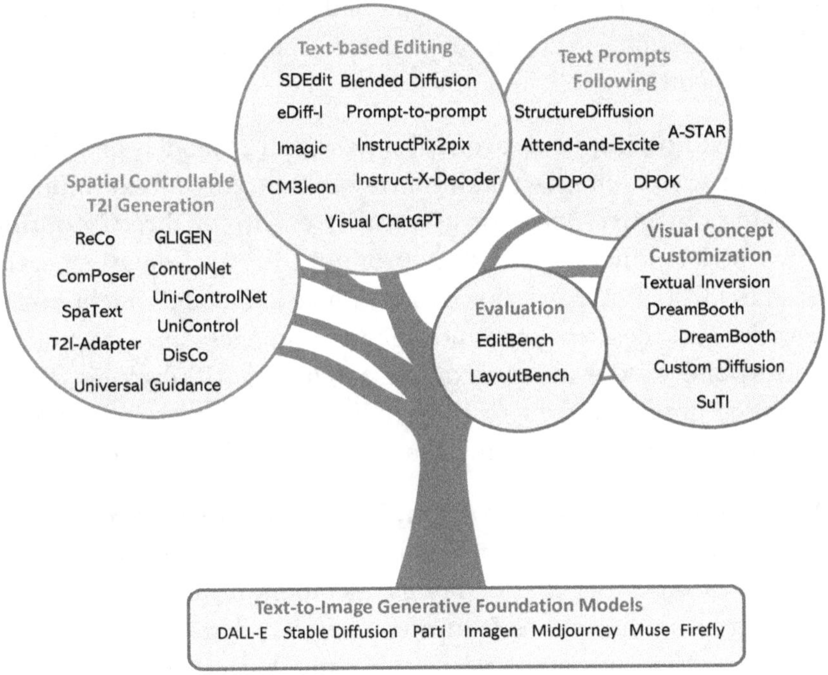

(b) Summary and categorization of papers on "Human Alignments in Visual Generation."

Figure 3.2: An overview of improving human intent alignments in T2I generation.

- **Spatial controllable T2I generation.** Text serves as a powerful medium for human-computer interaction, making it a focal point in conditional visual generation. However, text alone falls short in providing precise spatial references, such as specifying open-ended descriptions for arbitrary image regions with precise spatial configurations. Spatial controllable T2I generation [252], [511], [562] aims to combine text inputs with other conditions for better controllability, thereby facilitating users to generate the desired images.

- **Text-based image editing.** Editing is another important means for acquiring human-intended visual content. Users might possess near-perfect images, whether generated by a model or naturally captured by a camera, but these might require specific adjustments to meet their intent. Editing has diverse objectives, ranging from locally modifying an object to globally adjusting the image style. Text-based image editing [36] explores effective ways to create a versatile editing tool.

- **Better following text prompts.** Despite T2I models being trained to reconstruct images conditioned on the paired text input, the training objective does not necessarily ensure or directly optimize for a strict adherence to text prompts during image generation. Studies [376], [531] have shown that vanilla T2I models might overlook certain text descriptions and generate images that do not fully correspond to the input text. Research [31], [120] along this line explores improvements to have T2I models better following text prompts, thereby facilitating the easier use of T2I models.

- **Visual concept customization.** Incorporating visual concepts into textual inputs is crucial for various applications, such as generating images of one's pet dog or family members in diverse settings, or crafting visual narratives featuring a specific character. These visual elements often encompass intricate details that are difficult to articulate in words. Alternatively, studies [68], [380] explore if T2I models can be customized to draw those visual concepts with specialized token embeddings or conditioned images.

Before introducing the alignment works in detail, we first review the basics of text-to-image generation in the next section.

3.1.2 Text-to-Image Generation

T2I generation aims to generate images that are not only of high visual quality but also semantically correspond to the input text. T2I models are usually trained with image-text pairs, where text is taken as input conditions, with the paired image being the targeted output. Abstracted from the wide range of T2I models shown in Figure 3.3, we give a high-level overview of the representative image generation techniques.

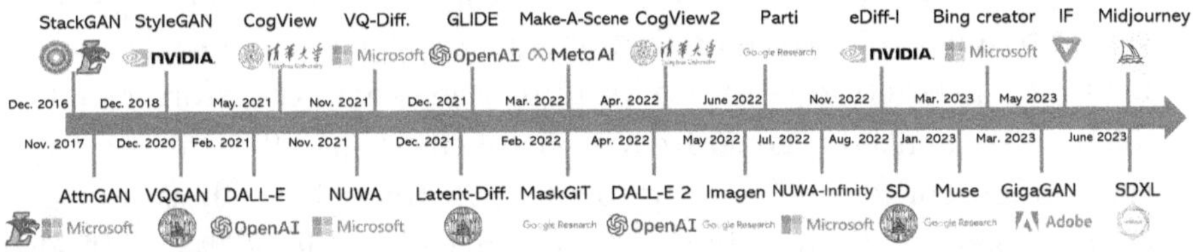

Figure 3.3: An overview of representative text-to-image generation models until July 2023.

- **Generative adversarial networks (GAN).** GANs [88], [143], [199] consist of two key components: a generator and a discriminator. The generator is tasked with creating synthetic images from random noise inputs, and it is trained to adjust these noise inputs based on input text conditions to generate semantically relevant images. In this adversarial process, the discriminator competes with the generator, attempting to differentiate between the synthetically generated images and real ones, thus guiding the generator to improve its image creation capabilities.

- **Variational autoencoder (VAE).** Variational Autoencoder (VAE) [204], [330], [436] is a probabilistic model that can generate images by employing paired encoder and decoder network modules. The encoder network optimizes the encoding of an image into a latent representation, while the decoder refines the process of converting the sampled latent representations back into a new image. VAEs are trained by minimizing the reconstruction error between the original and decoded images, while regularizing the encoded latent space using the Kullback-Leibler (KL) divergence. Vector Quantised-VAE (VQ-VAE) [330] further improves VAEs by leveraging the discrete latent space through vector quantization, enabling improved reconstruction quality and generative capabilities.

- **Discrete image token prediction.** At the core of this approach lies a combination of a paired image tokenizer and detokenizer, like Vector Quantized Generative Adversarial Networks (VQGAN) [112], which efficiently transform continuous visual signals into a finite set of discrete tokens. In this way, the image generation problem is converted to a discrete token prediction task. A widely employed strategy for token prediction is to use an auto-regressive Transformer [368], [531] to sequentially generate visual tokens, typically starting from the top left corner and moving row-by-row towards the bottom right, conditioned on the text inputs. Alternatively, studies [48], [49] also explore the parallel decoding to speed up the token prediction process. Finally, the predicted visual tokens are detokenized, culminating in the final image prediction.
- **Diffusion model.** Diffusion models [173], [407], [409] employ stochastic differential equations to evolve random noises into images. A diffusion model works by initiating the process with a completely random image, and then gradually refining it over multiple iterations in a denoising process. Each iteration predicts and subsequently removes an element of noise, leading to a continuous evolution of the image, conditioned on the input texts.

We use Stable Diffusion (SD) [376] as an example to explain in detail how T2I models work. We choose this model for a variety of reasons. Firstly, SD is one of the most widely used open-source T2I models, which makes it a solid foundation for many alignment techniques we discuss in this section. Additionally, as a diffusion-based generation model, it serves as an excellent case study for introducing diffusion models. Finally, its cross-attention-based image-text fusion mechanism is a classic example of various text-conditioned methods, such as auto-regressive T2I generation [531], helping us gain an in-depth understanding of the image-text interaction in T2I generation.

Stable Diffusion (SD),[1] and its academic version latent diffusion [376], contains mainly three modules, *i.e.*, an image VAE, a denoising U-Net, and a condition encoder, as shown in the left, center, and right part of Figure 3.4, respectively. We will introduce each module and the inference flow for image generation, following the notations in the original latent diffusion paper [376].

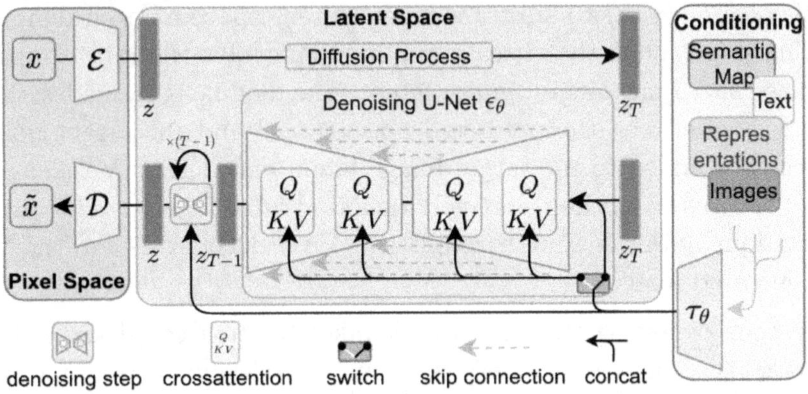

Figure 3.4: An overview of the latent diffusion model architecture. Image credit: [376].

- **VAE.** As introduced in the image generation technique overview, the VAE module contains a paired encoder $\mathcal{E}$ and decoder $\mathcal{D}$, trained to encode RGB image x into a latent random variable z and then decode the latent to best reconstruct the image. Given an RGB image $x \in \mathbb{R}^{H \times W \times 3}$, the encoder $\mathcal{E}$ encodes it into a continuous latent representation $z \in \mathbb{R}^{h \times w \times c}$. With the parameters of $H = W = 512$, $h = w = 64$, and $c = 4$ in SD, latent z is 48 times smaller than image x, thereby significantly improving the computational efficiency by performing the denoising process in this compressed compact latent space.

[1]We use Stable Diffusion v1 for the introduction. Later versions such as SD2 and SDXL share the same method but may have different detailed model configurations, such as a larger text encoder, U-Net, and latent dimension.

- **Text encoder.** SD is a conditional image generation model, where the input text condition is encoded using a condition encoder τ. Specifically, SD uses the ViT-L/14 CLIP text encoder [360] that encodes the tokenized input text query y into text feature $\tau(y) \in \mathbb{R}^{N \times d_\tau}$, where the maximum length N is 77 and text feature dimension d_τ is 768.

- **Denoising U-Net.** The denoising U-Net is the core module for the diffusion image generation process. The module is trained to predict the noise $\hat{\epsilon}(z_t, t)$ to subtract in the latent space at each denoising timestep t, such that it can step-by-step evolve the initial random noise into a meaningful image latent. The module is trained with the L2 loss between the predicted noise $\hat{\epsilon}(z_t, t)$ and the target noise ϵ, which is added to the target image latent encoded by VAE encoder $\mathcal{E}$. At inference, the iteratively denoised latent z, started from the random noise, is sent through the VAE decoder $\mathcal{D}$ for the final generated image.

  In each denoising step, the U-Net takes the text condition as input to generate images that are semantically relevant to the text query. We next detail how the visual stream $z \in \mathbb{R}^{h \times w \times c}$ interacts with the text stream $\tau(y) \in \mathbb{R}^{N \times d_\tau}$. The denoising U-Net, similar to a classic U-Net [284], [377], consists of a series of spatial downsampling and upsampling blocks with skip connections in between. In SD's U-Net, each down/upsampling block has a cross-attention layer and a 2D convolutional down/upsampling layer. Each block takes the visual latent feature, text feature, and denoising step as input and generates the next visual latent feature. The image-text interaction happens in the image-text cross-attention layer.

$$\text{Attention}(Q, K, V) = \text{softmax}\left(\frac{QK^T}{\sqrt{d}}\right) \cdot V, \tag{3.1}$$

  where K, V are projected from the text stream $\tau(y)$ and Q is projected from the visual stream z to share the same hidden dimension d. Therefore, the softmax between Q and K produces an attention map M of size $(hw \times d) \cdot (N \times d)^T = hw \times N$. The cross-attention map M indicates the fine-grained image-text interaction among each one of the N text words in all spatial positions hw. The attention map M then products V to yield the output of a down/upsampling block.

Recent advancements offer alternatives to SD, such as Diffusion Transformers (DiTs) [342], which represent an innovative category of diffusion models utilizing transformer architecture. These models diverge from the standard U-Net backbone, instead employing transformers that process latent patches. Residual Quantization [220] has also demonstrated its effectiveness in autoregressive image generation. Additionally, the architecture of DALL-E 2 [366] merits attention. It incorporates a prior model that generates CLIP image embeddings from text embeddings, followed by a diffusion U-Net conditioned on these CLIP embeddings. This structure, distinct from LDM or Imagen, proves particularly effective for tasks like stylization or reference-guided editing.

3.2 Spatial Controllable Generation

T2I generation takes open-ended text for users to describe their intended images. However, text alone is ineffective in certain descriptions, such as spatial referring. Studies in spatial controllable T2I generation explore extending T2I models to take extra spatial input conditions to guide image generation process.

We categorize related studies into three topics. (*i*) We start with works [13], [84], [252], [511] that extend the image-level text description in vanilla T2I models to the region-grounded text description, such that open-ended text descriptions can precisely operate on a particularly spatial region. (*ii*) We then extend from boxes to dense spatial conditions represented as 2D arrays, such as segmentation masks, edge maps, depth maps, key points. We review representative works ControlNet [562] and many others [320], [358], [547], [577]. (*iii*) The previous two threads of work require finetuning T2I models to understand the extended spatial condition. We next review techniques of inference-time guidance [25], [60] that achieve spatial control without model finetuning.

Region-controlled T2I generation. Large-scale T2I models have demonstrated remarkable efficacy in generating high-resolution images. However, the models lack spatial controllability, *e.g.*, precisely specifying content in a specified area using a free-form text description. This limitation motivates the studies on region-controlled T2I generation. As shown in the left side of Figure 3.5, these studies explore the extra input condition of open-ended text descriptions on arbitrary regions (*i.e.*, region-controlled text), augmenting the global image description in T2I models. This new input condition requires T2I models to understand spatial inputs, and associate them with grounded texts.

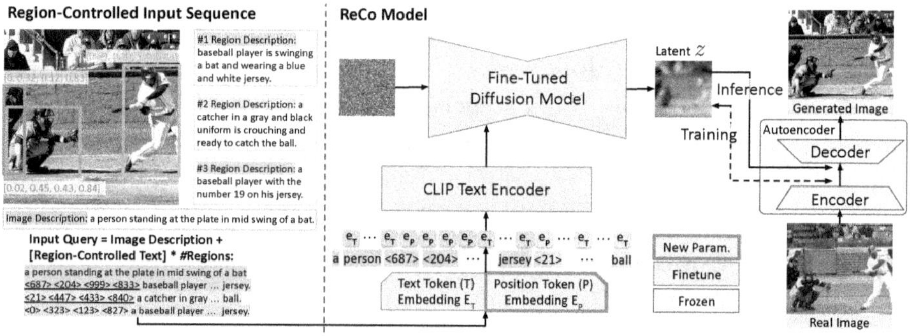

Figure 3.5: Overview of the ReCo model architecture. Image credit: [511].

ReCo [511] is among the most representative works along this direction. The core idea is to extend the text vocabulary of the text encoder $\mathcal{E}$ and arrange different tokens to represent the grounded text inputs. The study augments text tokens understood using pre-trained T2I models with an extra set of position tokens, which represent the quantized spatial coordinates. As shown in Figure 3.5, the position tokens (*e.g.*, <687>, <204>, <999>, <833>) are seamlessly mixed with the text tokens and operate as a spatial modifier, indicating that the text to follow only operates on the specified spatial region, such as the "baseball player ... jersey." The pre-trained T2I model is then finetuned to support such a new input interface, thereby facilitating region-controlled T2I generation.

Shared by other approaches along this direction, ReCo discusses several advantages of region-controlled T2I generation in improving the alignment with human intents. (*i*) The grounded texts provide an extra input condition that allows users to specify the desired image easily, *i.e.*, having a free-form regional description precisely at a specific location. The box token and the input sequence design allow users to generate grounded text with the same user interface as query a T2I model with text, making the extension easy to use. (*ii*) The additional region-level controlled texts help better generate images with correct object count, spatial relationship, and region attributes such as color/size, which may otherwise confuse the vanilla T2I model [376]. (*iii*) Studies also observe a better image generation quality, with the conjecture that the grounded text provides object-level image-text association and therefore simplifies the learning process.

GLIGEN [252] is another representative work. Alternate to generating grounded descriptions through the expansion of input tokens and finetuning the entire T2I model, GLIGEN uses a plug-and-play recipe: freezing the original T2I model and training extra gated self-attention layers to learn the new grounding skills. The grounding tokens carry two types of information: the semantic representation of text words that need to be grounded in and their spatial configurations. These grounding tokens are then added to the pre-trained T2I model via a newly added gated self-attention layer, with all remaining pre-trained parameters frozen. This layer is equipped with a gating parameter, which is initialized to zero, allowing the pre-trained model to incrementally incorporate the grounded text inputs. GLIGEN facilitates various types of grounded controls, including bounding box grounding, keypoint grounding, image prompting, as well as other types of spatially-aligned dense conditions.

T2I generation with dense conditions. In addition to spatial coordinates, there exist other spatial conditions often represented as 2D arrays, such as segmentation masks, edge maps and depth maps. ControlNet [562] is a prominent example of incorporating these dense spatial controls into T2I models. ControlNet is built upon Stable Diffusion, and introduces an additional trainable ControlNet branch that adds an extra input condition to the text prompt. This extra condition can be a canny edge map, hough line, HED boundary, under sketching, human pose maps, segmentation masks, depth images, normal maps, or line drawing, each enabled with its distinct model copy. The added branch is initialized from the pre-trained downsampling blocks in the SD's U-Net. This branch takes the added visual latent and the extra dense condition as input. Before combining input dense conditions with visual latent in the input and merging the ControlNet branch's output back to SD's upsampling blocks, there is a unique zero-initialized 1×1 convolutional layer. This layer serves as a gated connector to gradually inject the extra condition into the pre-trained Stable Diffusion model. With the extra dense spatial control, ControlNet provides an effective channel of generation controllability. Follow-up studies such as Uni-ControlNet [577] and UniControl [358] further improve ControlNet by unifying the input condition, such that a single model can understand multiple input condition types or even take a combination of two conditions. Examples of the dense controls and the corresponding generated images are shown in Figure 3.6. Moreover, Disco [453] exemplifies the efficiency of ControlNet in the generation of human dancing videos, which aims to generate videos with controllable elements such as human subjects, video backgrounds, and motion pose sequences. The study successfully separates the background and human pose conditions, which are fed into two distinct branches of ControlNet, which condition on image frames and pose maps, respectively. This disentanglement of control from all three conditions allows Disco to accomplish high fidelity in both the human foregrounds and backgrounds. More importantly, it enables the arbitrary compositionality of human subjects, backgrounds, and dance movements.

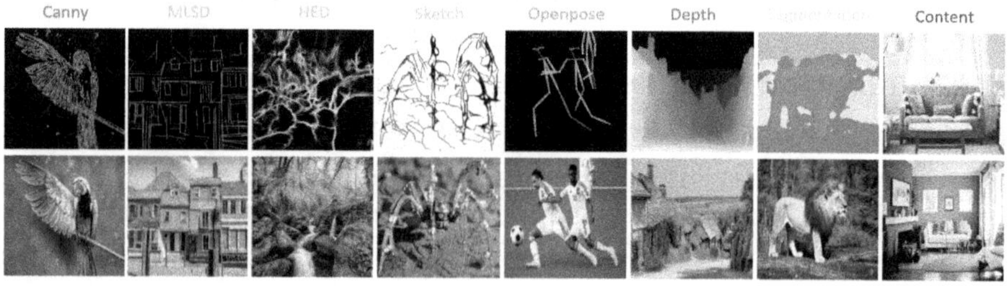

Figure 3.6: Examples of the dense controls and the corresponding generated images. Image credit: [577].

Inference-time spatial guidance. The aforementioned works require model training, either the T2I models or additional modules to understand the extra spatial conditions. Alternatively, studies [25], [60] explore providing the inference-time spatial guidance to T2I models without extra model training. The core idea is similar to classifier guidance [99], which takes a discriminator loss to guide the diffusion process as follows:

$$\hat{\epsilon}'(z_t, t) = \hat{\epsilon}(z_t, t) + s(t) \cdot \nabla_{z_t} \ell(c, f(\hat{z}_0)). \tag{3.2}$$

Taking spatial control as an example, the discriminator can be a Faster-RCNN object detector [373] indicated by f, which operates on the intermediate estimated image $\hat{z}_0$, and compute the object detection loss ℓ with the desired layout c, to guide the generation $\hat{\epsilon}(z_t, t)$. $s(t)$ is the guidance strength. This approach enables the spatial control in T2I generation without extra training, with qualitative results shown in Figure 3.7. However, it may not yield results as precise as those from finetuning methods [252], [511], [562].

(a) A headshot of a woman with a dog in winter. (b) a headshot of a woman with a dog on beach.

Figure 3.7: Qualitative results of inference-time spatial guidance. Image credit: [25].

Summary and trends. Early research on spatial controllable generation, such as layout-to-image and mask-to-image generation, was often treated in parallel with T2I generation. However, with the emergence of advanced large-scale T2I models, recent studies, as discussed in this subsection, are now leaning towards integrating spatial conditions with textual conditions. We identify two primary trends in integrating spatial conditions into T2I models. First, region-controllable T2I generation, such as ReCo, merges spatial coordinate control into text inputs by enlarging the text vocabulary with position tokens. Second, studies extended from ControlNet integrate an additional "image-like" condition to T2I frameworks, thereby capturing a broad spectrum of dense conditions. Moving forward, T2I models may have a finetuning stage that allows them to comprehend both image and text inputs. In such a scenario, box coordinates could be incorporated through text, while dense controls could be provided as image inputs. We will explore and elaborate on this idea in Section 3.5.

3.3 Text-based Editing

Text-to-image editing synthesizes new images from a given image and input text descriptions. Instead of producing an image entirely from scratch, users might already possess a satisfactory starting point; this could be an image previously generated from a T2I model or a natural image. The objective is to retain the majority of the visual content, only modifying specific components. This could involve altering a local object or the overall image style to precisely match the user's intentions. This text-based editing approach offers users a tool to generate fresh images based on a predecessor, playing a crucial role in creating visual content that accurately follows human intent.

There are various definitions and task setups in text-based editing. We introduce the following representative threads. (*i*) One classic editing scenario is to change a local image region, such as removing or changing an object or adding an object in a certain region. Spatially manipulating the latent in image generation according to the user-generated masks is a simple but effective method [12], [14], [306]. Studies [22], [170] also show that manipulating the image-text cross-attention mask is effective for spatial editing. (*ii*) Extended from spatial editing where the language inputs describe the desired appearance in the spatial region, language can also be used as editing instruction to tell the machine what to do [36], [201], such as "change object A in the image to object B." (*iii*) Instead of extending a single T2I model for editing, editing systems [477] integrate different specialized modules such as segmentation models [206], [601] and large language models [37], [334].

Diffusion process manipulations. The multi-step denoising process in diffusion image generation naturally supports a certain extent of image editing. Stochastic Differential Editing (SDEdit) [306] shows that first adding noises to the input image to edit and then subsequently denoising the sample, could produce a meaningful edit. Blended Latent Diffusion [12] shows that the diffusion process manipulation can achieve local object editing with a user-generated object mask m_{latent}. In each diffusion step, the latent z is a spatial blend of the foreground and background latent: $z = z_{fg} \odot m_{latent} + z_{bg} \odot (1 - m_{latent})$, where z_{fg} is the edited object generated from the text description and z_{bg} is the original background image with noises added.

However, there are certain limitations on blending spatial latents. Firstly, it may not always be feasible to require human-generated masks. Secondly, the generation process can sometimes result in artifacts at the edges. Instead of simply blending the latent in a spatial manner, researchers delve into image-text cross-attention maps to unearth clues for object editing. Specifically, Prompt2Prompt [170] discovers that cross-attention layers control the interaction among visual regions and text words. Based on this observation, the study enables three types of editing for images generated by a diffusion T2I model, including word swap, adding new phrases, and attention re-weighting, each of which is enabled with corresponding manipulation on the image-text cross-attention map. Specifically, the Prompt2Prompt tracks both cross-attention maps generated by the original prompt (namely M_t) and the edited prompt (namely M_t^*), and merges the attention maps with pre-defined rules into the new attention maps $\widehat{M_t}$, which is used for latent computing. For example, while adding a new phrase, attention map $\widehat{M_t}$ remains unaltered for words present in the original prompt. It only incorporates the modified attention maps M_t^* for words that did not exist in the original prompt. Qualitative results of the edits are shown in Figure 3.8.

Figure 3.8: Three types of editing (word swap, adding new phrases, attention re-weighting) on synthetically generated images, enabled by attention map manipulation. Image credit: [170].

Going beyond editing synthetically generated images, Imagic [201] explores editing real natural images. The core idea is to represent the image to be edited as text embedding, and blend this embedding with the target text embedding describing the desired image. This blend ensures that the resulting image retains elements from the original while aligning with the aesthetics detailed in the target textual prompt. In practice, test-time finetuning is needed to generate high-quality images.

Text instruction editing. Instead of repeating the visual contents of the image to edit in the text prompts, it might be more efficient for users to directly specify editing instructions using language, such as "swap sunflowers with roses" in Figure 3.9. The desired text instruction editing model should work on both model-generated and natural images, and across different types of editing instructions.

Figure 3.9: Examples of text instruction editing. Image credit: [36].

InstructPix2Pix [36] is designed to accept an image and a text editing instruction to produce an edited version of the input image. The goal is to train an image-to-image model that can understand such editing text instructions. To achieve this, T2I models can be adapted to accept the additional image input by incorporating more input channels into the SD's convolutional layer. The main challenge is how to generate paired editing data. As shown in Figure 3.10, InstructPix2Pix [36] proposes to use a LMM [37] to generate a pair of an editing instruction and an edited caption from the original input caption, *e.g.*, "have her ride a dragon," "photograph of a girl riding a dragon," and "photograph of a girl riding a horse." The study then uses Prompt2Prompt [170] to convert the original and edited caption pair to a pair of images before and after editing, corresponding to the GPT-generated editing instruction. The study generates over 450K samples to train the editing model. This data generation method has also been adopted in subsequent research, such as CM3Leon [133] for training general-purpose image-text-to-image models.

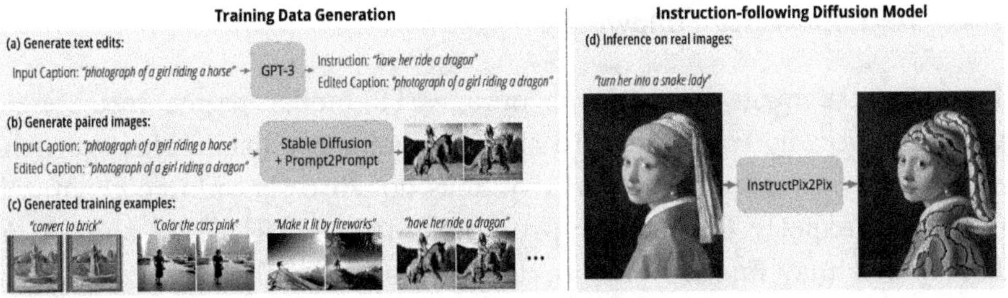

Figure 3.10: The editing data generation pipeline proposed in InstructPix2Pix. Image credit: [36].

Editing with external pre-trained models. Furthermore, recent studies show the efficacy of incorporating external language and vision models for editing, as opposed to relying solely on a single model. Advancements in generalist segmentation models, such as SAM [206] and SEEM [601], have paved the way for using segmentation models to ground the region for text-based editing. Representative works include Instruct X-Decoder [599], Grounded SAM inpainting [271], Inpaint anything [536], etc.. Another emerging trend is the allocation of various generation and editing tools through LMM. Studies such as VisualChatGPT [477] can solve complicated visual editing that requires the collaboration of multiple generation and editing models in multiple steps.

Summary and trends. Text-based editing models have made significant progress in their capabilities, leading to improved editing quality, expanded domain coverage, and more flexible user interface. For example, early studies require user-generated masks for object editing, while recent models can work on synthetically generated images without mask inputs, or even directly understand general text editing instructions. As we look to the future, we anticipate an all-encompassing generative foundation model that is capable of processing both image and text inputs. Within this framework, editing instructions would be a specialized form of text input, seamlessly integrated with the image description in T2I generation.

3.4 Text Prompts Following

Training with image-text pairs encourages T2I models to generate images that semantically correspond to the input text condition. However, the image generation training objective does not directly enforce generated images to exactly follow text prompts. Studies [51], [120] show that T2I models may fail to follow text prompts, especially when the image description becomes complicated. For example, certain noun phrases may get omitted, attributes may apply to incorrect objects, and generated images may have the wrong object count, relationship, styles, etc.. These limitations motivate work on improving T2I models to better follow text prompts.

The related literature can be broadly categorized into two main groups. (*i*) *Inference-time manipulation.* In the inference stage, the latent and attention adjustment [1], [51], [120], [267] design various methods to redistribute the visual latent or image-text cross-attention, such that all noun phrases in the text prompts are represented in the generated image. (*ii*) *Alignment tuning.* An extra model learning stage is learned [31], [115], typically with the image-text similarity as rewards, such that the tuned T2I model can better follow text prompts.

Inference-time manipulation. Training with image-text pairs does not guarantee that T2I models consistently adhere to the text prompts. There can be multiple discrepancies, particularly when the text descriptions are lengthy and intricate. For instance, T2I models may apply attributes to the wrong entity or miss certain objects, as shown in Figure 3.11. Intuitively, parsing the text query at inference time and explicitly enforcing T2I models to pay closer attention to each noun phrase may generate images that better follow text prompts.

Building upon this intuition, StructureDiffusion [120] employs a parsing tree to extract noun phrases and the text prompt's linguistic structure. The study then enforces the model to "look at" all extracted noun phrases. This is implemented by modifying SD's cross-attention mechanism introduced in (3.1), written as $O = M \cdot V$ where M is the softmax cross-attention map. Instead of producing M with the sentence feature V, which may result in words getting overlooked, the study computes the $O = \frac{1}{k+1}\Sigma_{i=0}^{k}(M \cdot V_i)$, where V_0 is the sentence

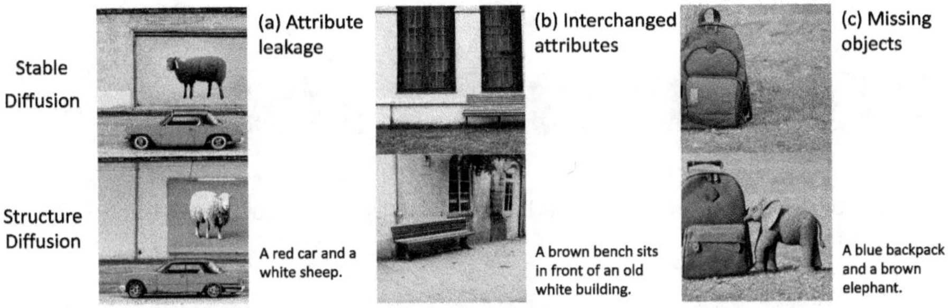

Figure 3.11: Failure cases of vanilla T2I model in text prompt following. Image credit: [120].

feature V, and $V_i, i = 1, \ldots, k$ is the phrase feature in the parsing tree. This approach ensures that the visual stream maintains a balanced attention across all identified noun phrases, fostering more accurate image generation.

Motivated by the same objective to ensure that no object is overlooked, Attend-and-Excite [51] manipulates the attention map. As shown in the right side equations in Figure 3.12, a regularization loss ℓ is computed to amplify the maximal attention towards the most neglected subject token:

$$\ell = \max_{n=1,\ldots,N_{\text{sub}}} (1 - max\ G(M_t^n))$$

where G is a Gaussian kernel to smooth the attention map and N_{sub} is the number of subject tokens. The loss is then used to update the latent z_t at inference time:

$$z_t' = z_t - \alpha \nabla_{z_t} \ell,$$

where α is a scalar for the step size. Results show that this inference-time guidance enables T2I models to focus more on objects described in the text prompt, resulting in superior image generation. Follow-up studies [1] further refine the guidance loss to optimize prompt-following performance.

Model tuning to follow text prompt. Instead of inference-time manipulation, one may wonder if we can refine a pre-trained T2I model to better follow text prompts. One promising way to achieve this is via reinforcement learning, using image-text similarity as reward instead

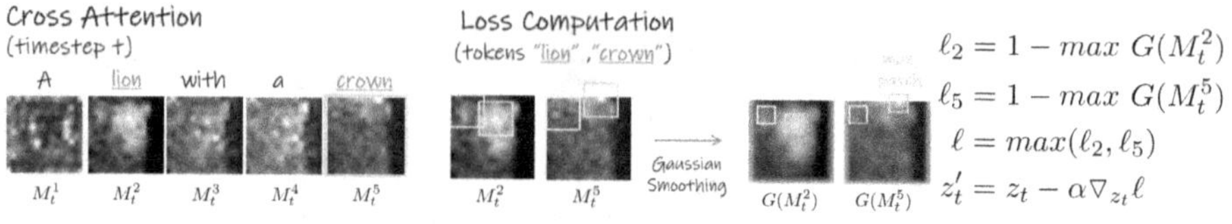

Figure 3.12: Inference time guidance proposed in Attend-and-Excite. Image credit: [51].

of the image generation objective used in the main T2I training. This allows the model to be optimized towards a better image-text alignment.

One work along this direction is the denoising diffusion policy optimization (DDPO) [31], with the tuning framework shown in Figure 3.13. Specifically, a vision-language model [228] is used to convert the generated image into a text description. This generated caption is compared with the input text prompt, deriving a similarity reward through the use of BERTScore [568]. The similarity reward is then used to finetune the pre-trained T2I model, such that the model can better follow the text prompts. The bottom of Figure 3.13 shows the progression of the generated sample during this similarity-based training. Further, it is worth noting that other human intent may also be formatted as rewards for model tuning, such as compressibility, aesthetic quality, etc.

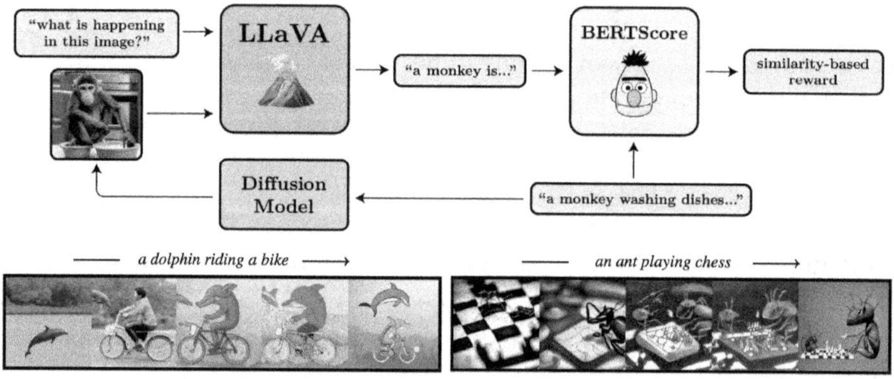

Figure 3.13: DDPO with vision-language-model-based reward function for image-text alignment tuning. Image credit: [31].

Summary and trends. In this section, we present studies aimed at enhancing the capability of T2I models to better adhere to text prompts. Despite the good performance achieved by the inference-time manipulation, the alignment tuning provides a more intuitive user experience, eliminating the need for extra modifications. In parallel to instruction tuning in LLMs to align human intent for text generation, the TI2 model tuning shares a similar goal, but focuses on image generation. We foresee a similar paradigm emerging in the near future for generative T2I foundational model development. Specifically, the initial training phase still relies on the existing image generation objective on large-scale data, while the subsequent alignment-tuning phase enhances the model's prompt adherence and other aspects of human intent, such as diminishing harmful content. Current RL-based methods show potentials, but they typically zero in on a single optimization goal. Future research could extend these methods for more adaptable alignment tuning, amalgamated with features like accommodating diverse image and text scenarios.

3.5 Concept Customization

Though language is a powerful medium to express human intent, it is inefficient in comprehensively describing all details of a visual concept for reconstruction. For example, it is challenging to use texts to describe my pet dog or family members with sufficient details, so that they can be generated in different visual scenes. In such applications, directly extending T2I models to understand visual concepts via image inputs is a better option.

We examine relevant research on visual concept customization, which offers users the ability to generate these personalized concepts. (*i*) Pioneer studies [127], [380], [474] start with single-concept customization that involves test-time finetuning to encode multiple images of the visual concept into a new token embedding, such that the learned embedding can be used to refer to the concept during T2I generation. (*ii*) Multi-concept customization [11], [213] allows multiple concept tokens to be expanded from the T2I model's token vocabulary, enabling multiple concepts to interact with each other and the remaining visual scene during generation. (*iii*) Test-time finetuning requires users to tune T2I models for each new concept to customize. To simplify the usage, studies [68], [69], [398], [502] explore customization without test-time finetuning and use a unified finetuning stage to extend T2I models for accepting image condition inputs. The models take images of the visual concept as an extra input condition, and generate images with the visual concept following the text descriptions.

Single-concept customization. The goal of visual concept customization is to enable T2I models to comprehend additional visual concepts tailored to very specific cases. The problem setup, studied in Textual Inversion [127], involves translating visual concepts from a handful of images into unique token embeddings. As illustrated in the left side of Figure 3.14, the T2I model processes four images of a distinct dog breed, subsequently learning the embedding for a new token, denoted as [V]. This [V] token can be used as a text token to represent this specific dog. The [V] token can be seamlessly integrated with other textual descriptions to render the specific dog in various contexts, such as swimming, in a bucket, and getting a haircut.

Figure 3.14: The problem setup and result visualization of the visual concept customization task. Image credit: [380].

Textual Inversion [127] learns the [V] token embedding via prefix tuning, *i.e.*, freezing all T2I model's parameters and training the [V] token embedding to generate the input images. Later studies observe that tuning more model parameters leads to significantly better image generation quality. However, adjusting only the input image may lead to the risk of overfitting the T2I model for a particular concept, and losing the capability to generate diverse images. For instance, the model might become unable to generate various dog types. To address this,

Dreambooth [380] proposes the class-specific prior preservation loss. Central to this approach is using the pre-trained T2I model to produce images of the same class as the targeted customization concept. The model is then jointly finetuned on both the input image (with the [V] token) and the model-generated images (without the [V] token). This ensures that the model can differentiate between the unique "[V] dog" and other general dogs it was initially trained, thus maintaining its overall T2I capability. Dreambooth then finetunes all T2I model parameters and achieves better image generation quality.

Multi-concept customization. Building on studies that focused on learning a single visual concept [V], recent research has delved into the possibility of integrating multiple visual concepts into a single Text-to-Image (T2I) model, represented as [V_1], [V_2], and so on. Custom Diffusion [213] employs a selective subset of model weights, specifically the key and value mappings from text to latent features in the cross-attention layers for concept customization, learned from multiple sets of concept images. The study facilitates the ability to embed multiple customized visual concepts in a single text prompt. Instead of learning from multiple sets of input images, Break-A-Scene [11] explores extracting multiple visual concepts in a single image. The study augments input images with segmentation masks to pinpoint the intended target concepts and subsequently transforms them into a series of concept embeddings denoted as [V_i].

Customization without test-time finetuning. While the concept customization studies, as described above, have achieved good visual quality, the necessity for test-time finetuning hinders its application in real-world settings. Most end users and application platforms lack the compute resources required for finetuning, not to mention the complexities of finetuning process. This naturally leads to the question: can we take concept images as input conditions, and achieve concept customization without finetuning?

The input/output format of the imagined system is similar to the retrieval-augmented generation [69], which aims to ease the image generation by conditioning on a retrieved similar image. The system supports extra image inputs that contain relevant information for the generation process. By altering the conditioning images during the training phase, the model can potentially achieve a broad in-context learning capability, producing images that align with the given input examples. In line with this framework, SuTI [68] trains a single model to imitate the finetuned subject-specific experts, and generates images conditioning on both text and subject input images, as shown in Figure 3.15. As a result, the trained model can perform in-context concept customization without test-time finetuning, and generalize to unseen subjects and descriptions. Another concurrent work, InstantBooth [398], also shows remarkable results in generating images that are not only aligned with language but also preserve identities, with a single forward pass.

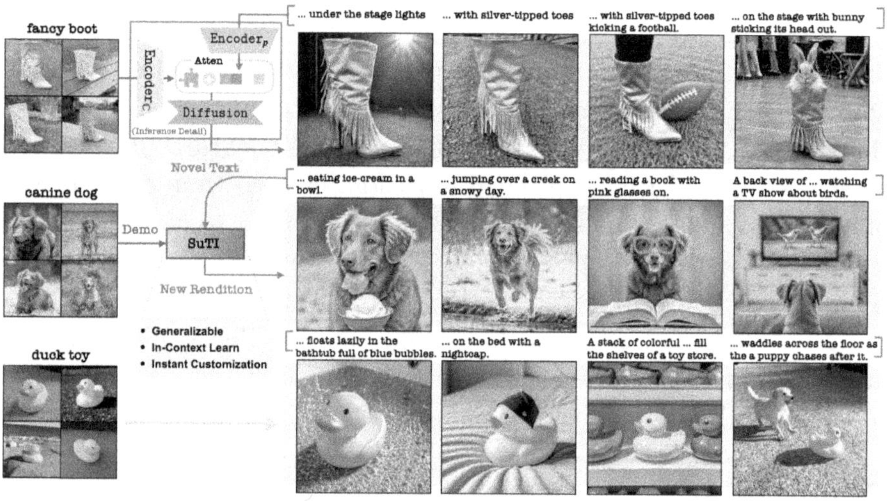

Figure 3.15: Illustration of in-context concept customization without test-time finetuning. Image credit: [68].

Summary and trends. The field of visual concept customization has advanced from finetuning embeddings during the testing stage, to directly performing in-context image generation with a frozen model. The in-context generation pipeline, which incorporates additional image inputs, shows remarkable potential in real-world applications. In this subsection, we have explored two applications of this approach: facilitating generation through the retrieval of pertinent images [69], and personalizing visual concepts by conditioning them on subject images [68], [398]. An intriguing direction is to unify the diverse uses of image inputs, directed by descriptive textual instructions. We elaborate on this idea in the following sub-section.

3.6 Trends: Unified Tuning for Human Alignments

In previous subsections, we presented the literature related to tuning T2I models to more accurately align with human intent. This includes enhancing spatial controllability, editing existing images for improved alignment, more effectively following text prompts, and personalizing T2I models for new visual concepts. A trend observed across these subtopics is the shift towards integrated alignment solutions that require minimal problem-specific adjustments. Along this direction, we envision a future T2I model having a unified alignment tuning stage, which transforms a pre-trained T2I model into one that resonates more intimately with human intent. Such a model would seamlessly process both text and image inputs, generating the intended visual content without the need for multiple models tailored to different alignment challenges.

Drawing parallels to the established practice of human-alignment tuning in LLM development, we anticipate that the techniques reviewed in this section will merge into a holistic second-stage tuning for generative foundation model development. This alignment tuning phase serves two primary purposes. First, it extends the T2I's text inputs to incorporate interleaved image-text inputs, as illustrated in Figure 3.16. Second, it finetunes the base T2I model, which has been trained using image generation loss, by the employing data, loss, and rewards that aim to align with human expectations.

Unified image and text inputs. We begin with the discussion on interface unification. Specifically, we aim to evolve the textual inputs of T2I models into a multimodal interface that seamlessly integrates both image and text inputs. As shown in Figure 3.16, we consider three types of inputs to begin with: 11content text input" characterizes the visual scene to be produced; the "image input" accommodates dense 2D inputs such as images and dense conditions; and the "instruction text input" explains how the input content texts and images should be collectively composed as the condition for generation.

Related topics	Instruction text input	Content text input	Image input
T2I models (Sec. 3.1.2; SD)	None	Image description	None
Region-controlled T2I (Sec. 3.2; ReCo)	None	Image description + Box Tokens	None
T2I with dense conditions (Sec. 3.2; ControlNet)	None	Image description	Dense conditions (segmentation, edgemap, depth, keypoints, etc.)
Text instruction editing (Sec. 3.3; InstructPix2Pix)	Editing instruction ("change the dog's color to blue")	Contents for editing instruction	Image 
Concept customization (Sec. 3.5; SuTI)	Customization instruction ("generate a dog looks like this one")	Image description	Image 
Alignment tuned T2I models (Sec. 3.6)	Arbitrary instruction ("generate a dog looks like the left one but in blue")	Image description + Box Tokens ("in <687>, <204>, <999>, <833>")	Image or dense conditions 

Figure 3.16: Overview of the unified image and text input interface for human alignments, and its connection to previous sub-sections.

Vanilla T2I models, as shown in the first row of Figure 3.16, take the "content text input" of the image description and generate the corresponding image. For the spatial controllable generation in Section 3.2, the extra spatial condition can be specified via text inputs by expanding text words with extra box tokens, or via image input by feeding the dense spatial conditions as an image input. For the text-based editing in Section 3.3, we examine the efficacy of text instruction editing, a task that finetunes the T2I model to comprehend editing instruction texts that manipulate the image input, altering its pixel values accordingly. For visual concept customization in Section 3.5, the training-free models can now understand customization instructions to extract visual concepts from the image inputs, and combine the concept with context text inputs for image generation.

Incorporating the three elements of the input interface, the envisioned alignment-tuned T2I model can handle all previous tasks described in Sections 3.2-3.5. Its behavior is steered by specific text instructions that dictate how the image and text inputs should be jointly processed as the generation condition. Given the same image input, different text instructions can invoke different tasks: "generate a cat image with the same layout" for spatial control, "change the dog's color" for editing, "generate the same dog sleeping" for concept customization, and the arbitrary mixture of the existing modes. Achieving such a unified interface in generative foundational models may be possible through training on a consolidated dataset encompassing data from various tasks, drawing similarities to the success of supervised instruction tuning observed in LLMs. Furthermore, transitioning from processing a single image-text pair to handling interleaved image-text pairs could enable more intriguing capabilities like in-context visual demonstrations [417]. Another interesting direction is to build a generative model that is capable of generating any combination of output modalities, such as language, image, video, or audio, from any combination of input modalities, as demonstrated in Composable Diffusion (CoDi) [423].

Tuning with alignment-focused loss and rewards. In addition to the unified input interface, another noteworthy element deserving consideration is the alignment-focused loss and rewards. As mentioned in Section 3.4, the image generation loss based on image-text pairs enables models to produce images that match the target data distribution. Yet, it doesn't always perfectly align with human intent. This is reminiscent of the language model loss in LLM training, which necessitates a separate alignment tuning phase [338]. The recent success in supervised instruction tuning and reinforcement learning from human feedback methods [31] on image generation provides effective tools for similar alignment tuning in generative foundation models. An intriguing topic left for future exploration is how to balance the different target losses and rewards, such as jointly optimizing for higher aesthetic scores, better image-text alignment, fewer harmful contents, stronger instruction adherence, along with many other desired properties.

Text-to-any-modality generation. The research theme of visual generation has been generalized to modalities beyond 2D images, such as synthesis of video, 3D and speech. (*i*) *Text-to-video (T2V).* T2V synthesis refers to the creation of videos from textual descriptions. This process involves producing a series of images that are coherent both in time and space. While T2I synthesis has reached a state of maturity,

T2V represents a relatively novel endeavor within the field. Recently, there has been significant advancement in T2V technology, marking a leap forward in the capability to generate dynamic visual content from textual input. Several recent T2V works include Gen-2 [134], MoonValley [314], Morph [316], Pika 1.0 [352], Stable Video Diffusion [32], [420] MagicVideo-V2 [455] *(ii) Text-to-3D.* The surge in the requirement for 3D digital content across various domains such as gaming, entertainment, architectural design, and robotics simulation is noteworthy. The process of crafting this 3D content necessitates a profound level of artistic flair and specialized training in 3D modeling, which is a time-intensive and strenuous endeavor. By integrating natural language techniques into the 3D content creation process, there is a potential to make this field more accessible to beginners and simultaneously enhance the productivity of seasoned artists. Several attempts have been on text-to-3D, including Dream Fields [191], Dream Fusion [355], Magic3D [256], zero-1-to-3 [268] and ProlificDreamer [466]. *(iii) Text-to-Speech.* Whisper [361] and its V2/V3 family are probably the most representative speech recognition models that are capable of transcribing and translating multiple languages with improved accuracy in accent and noise recognition. Trained on 680,000 hours of diverse audio data, it excels across various datasets, reducing errors by 50% compared to other models, although its performance varies with languages less represented in training. Whisper also forms the basis for a unified model encompassing broader sound recognition.

Fast inference of diffusion models. The increasing utilization of diffusion models in visual generation contexts has rendered the efficiency of their inference a crucial aspect. It has led to the emergence of a novel category of generative models, which are pivotal in distilling diffusion models to decrease the number of sampling steps. Several attempts in this domain include the Consistency Model [408], Rectified Flow [275], and UFOGen [499]. This progress is significant in facilitating multimodal generation. The current disparity in generation speeds between visual and textual modalities highlights the limitations of conventional autoregressive models like GPT for scalability, underscoring the need for more rapid decoding mechanisms. Illustratively, OpenAI's DALLE3 has transitioned to a decoder based on the Consistency Model. This shift underscores the evolving landscape in generative model technologies, particularly in enhancing efficiency and applicability across various modalities.

Closed-loop of multimodal content understanding and generation. As we look ahead, one promising avenue of research is the closed-loop integration of multimodal content understanding and generation. Preliminary studies have shown the benefit of using synthesized data to benefit generation from understanding [168], [222], and vice versa. An exciting prospect would be the development of an image-text-input, image-text-output foundational model for both understanding and generation tasks. The ideal balance in combining these two dimensions, and the most efficient approach to achieve it, are left for future explorations.

4

Unified Vision Models

In this section, we discuss the unification of vision models. We start with an overview of the challenges in the unification of vision models and the most recent efforts towards this goal in Section 4.1. What follows are detailed discussions on (*i*) how to transform closed-set models to open-set ones in Section 4.2; (*ii*) how to unify different granularities of vision tasks in Section 4.3; and (*iii*) how to build a more promptable interface for vision in Section 4.4. Finally, we summarize the section and discuss future trends in Section 4.5. Figure 4.1 depicts an AI model navigating through a fantastical world, where visual understanding and synthesis converge seamlessly.

4.1 Overview

Before describing general-purpose unified vision systems, we revisit how language models and natural language processing (NLP) have evolved in the past years. Before 2018, different NLP tasks are addressed with different task-specific models, such as translation [19], semantic parsing [29], summarization [5], and so on. With the emergence of the transformer architecture [437], language models for different NLP tasks are unified with a decoder-only architecture, e.g., the GPT models [37].

Figure 4.1: An AI with unified visual capability navigating through a fantastical world.

Afterwards, the GPT models learned using the next word prediction task are further finetuned to follow human instructions. This leads to ChatGPT,[1] which fundamentally changes our expectations on what AI systems can do. The evolution as depicted in Figure 1.2 motivates us to wonder whether we can build a general-purpose vision system in a similar manner.

Challenges. That computer vision tasks vary greatly presents a great challenge to build a unified vision model. First, vision tasks have different types of inputs, ranging from static images [381] to sequential videos [308], from pure vision inputs such as image dehazing [165] to multi-modality inputs that include e.g., vision and language [8]. Second, different granularities are required for different tasks, such as image-level tasks like image classification [166] and captioning [439], region-level tasks like object detection [140] and grounding [353], and pixel-level tasks like image segmentation [164], super-resolution [467], *etc.* As a result, the outputs of vision systems are also of different formats, such as spatial information like edges, boxes, and masks, semantic information like class labels, multi-label tags, or detailed descriptions. In addition to the challenges in modeling, there are also challenges with data. First,

[1]https://chat.openai.com/

the cost of annotation varies greatly among different types of labels. As shown in Figure 4.7, these labels are at different levels of granularity and semantic richness, ranging from whole images, regions (box annotations), to masks (pixel annotations). Second, it is in general much more costly to collect image data than text data. So, the scale of vision data is often much smaller than that of text corpora.

Towards a unified vision model. Despite these challenges, there is a growing interest in the computer vision community to develop a general-purpose, unified vision system, in particular for visual understanding tasks. As illustrated in Figure 4.2, we group these efforts in three categories:

- **Bridging vision and language**. By extending closed-set classification to open-set recognition, the contrastive language-image models like CLIP [360] demonstrate impressive zero-shot transferability for different vision tasks. These models learn the mapping between raw visual signals and rich semantics and can power various open-vocabulary vision recognition tasks [138], [147], [242], [582].
- **Unified multi-task modeling**. Traditional task-specific vision models are trained using task-specific data. It is often prohibitively expensive to develop a model for a new task. Thus, it is desirable to develop a unified vision model that can perform well across many vision tasks [64], [287], [509], [599].
- **LLM-like promptable interface**. LLMs can take different language and in-context prompts as inputs and produce user-desired outputs without finetuning. A general-purpose vision model should have possessed the same in-context learning capability to align the output to various user intents without changing its model parameters [23], [27], [206], [460], [601].

In what follows, we will elaborate the detailed techniques and methods in each category.

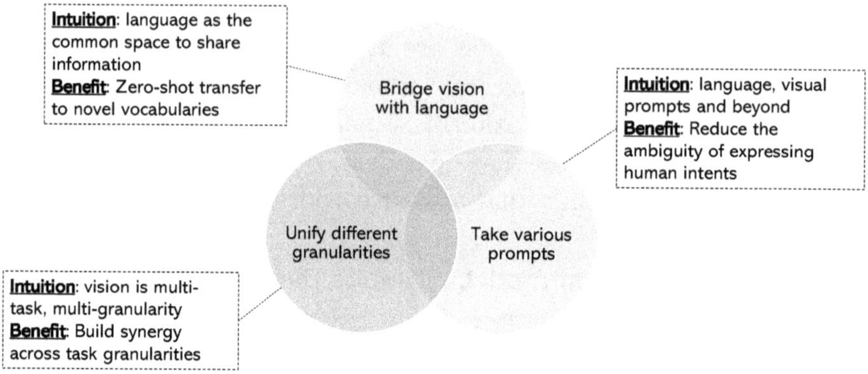

Figure 4.2: In NLP, we have witnessed a trend to build a unified model such as GPT-3 [37] and then the sophisticated Human-AI interaction system ChatGPT, which has ignited the interests of the whole community and society in AI. A natural question for computer vision (CV) is whether we can unify all different types of vision tasks such as image classification, object detection, segmentation and visual question answering, *etc.*, and likewise build an interaction interface between CV models and humans. Inspired by this, a lot of attempts have been made recently to crack the problem from different angles including but not limited to (a) making vision models open-set; (b) unifying different granularities; and (c) making the models more promptable.

4.2 From Closed-Set to Open-Set Models

Traditionally, visual recognition is formulated as a classification problem that maps raw visual data (e.g., images) to discrete text labels. For example, image classification predicts a label from a pre-defined close set for a whole image [94], and object detection identifies the objects, defined in a closed set, within an image [258]. However, such *closed-set* models can hardly transfer to other tasks where the closed set (or vocabulary) is insufficient. For example, it is difficult to apply an object detector trained using the Microsoft COCO object set[2] to detect Minecraft objects. Recently, CLIP [360] addresses the limitation of closed-set models by introducing a contrastive language-image pre-training method to train an *open-set* model. As illustrated in Figure 4.3 (a), instead of learning the mapping from input to labels, CLIP learns an aligned

[2]https://cocodataset.org/

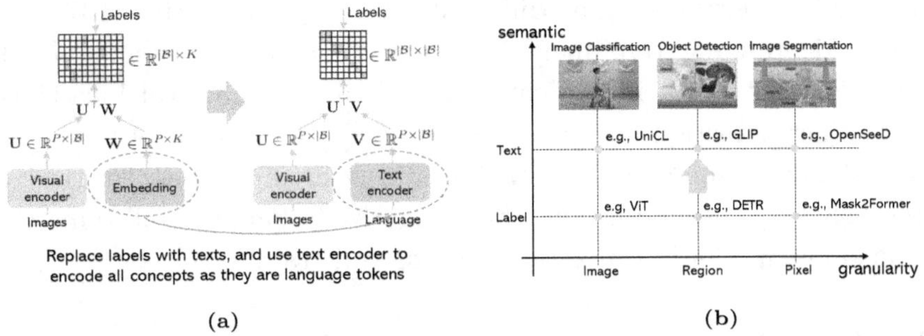

Figure 4.3: (a) As proposed in CLIP [360], replacing labels with textual descriptions and using a text encoder to encode them can feasibly convert closed-set problems to open-set ones. Image credit from [504]. (b) A number of works have been proposed to transform different computer vision tasks by replacing the label space with language space, such as UniCL [504], GLIP [242] and OpenSeeD [557].

visual-semantic space using hundreds of millions of image-text pairs. Mathematically, the traditional vision tasks optimize the log-likelihood of assigning label $y = c$ to an image, which is often represented as feature vector $u \in \mathcal{R}^P$:

$$\log \mathcal{P}(y = c|u) = \log \frac{\exp^{w_c \cdot u}}{\sum_{i=1}^{K} \exp^{w_i \cdot u}}, \tag{4.1}$$

where $w \in \mathcal{R}^{K \times P}$ is a projection matrix. Instead of using a pre-determined project matrix w, the CLIP method uses a text encoder Enc_{text} for the projection:

$$v_i = Enc_{text}(Concept_i), \tag{4.2}$$

where v plays the role of w in Equation (4.1). The reason why a text encoder can help achieve open-set recognition is that all textual concepts are embedded in the same feature space through large-scale pre-training, and the feature distributions are coherent to the semantic meanings without the need of a pre-defined vocabulary. As such, the aligned visual-semantic space can be easily transferred to a wide range of image recognition tasks in a zero-shot manner. Please refer to Section 2 for a detailed discussion. In the following, we focus our discussion on the region-level and pixel-level models.

After the release of the CLIP model [360], a number of open-set vision models have been developed using large amounts of text-image pairs for visual understanding at different levels of granularity [137], [242], [504], [557], ranging from image-level tasks (*e.g.*, image classification [94], image-text retrieval, image captioning [76]), region-level localization (*e.g.*, object detection and phrase grounding [353]), to pixel-level grouping tasks (*e.g.*, image segmentation and referring segmentation [160], [205], [284]). These models can be categorized along the following three dimensions: model initialization, design and training.

Model initialization. There are different initialization methods for open-set model training.

- **CLIP initialized.** Many recent open-set models are trained by using a pre-trained model such as CLIP for initialization since a pre-trained model already provides a well-aligned (but often coarse-grained) visual-semantic feature space. For example, OVR-CNN [543] and RegionCLIP [582] use a CLIP-style pre-trained ResNet [166] as the vision encoder and a pre-trained RPN [373] to extract regional features. Likewise, MaskCLIP [584] and FreeSeg [359] exploit the CLIP model to extract dense labels for pixels. FC-CLIP [535] uses a frozen convolution network ConvNeXt [282] in CLIP to encode input images of various resolutions.
- **CLIP augmented.** Instead of initializing a model with CLIP parameters, other methods initialize the model parameters as usual (e.g., setting random values to model parameters), but use the pre-trained CLIP to help model training. For example, ViLD [147] augments the model with aligned CLIP features via knowledge distillation. MaskCLIP [101] and Mask-Adapted CLIP [253] rely on the pre-trained CLIP model to provide features and scores, respectively, during the course of model training.
- Other works learn a visual-semantic feature space using supervised pre-trained models or from scratch. For example, GLIP [242] and OpenSeeD [557] use a pre-trained BERT model [98] and the CLIP text encoder, respectively, and use a vision backbone pre-trained on ImageNet for image encoding. Though these separately pre-trained image and text encoders do not explicitly learn the alignment between

image and language, it turns out that these models still give good representations for images and texts, and are instrumental to efficient model training. Differently, GroupViT [494] is trained jointly using an open-set semantic segmentation task and a global image-text alignment task from scratch. ODISE [495] exploits pre-trained Stable Diffusion models [376] to extract compact masks.

Model design. Open-set models can be either multi-stage or end-to-end.

- **Two-stage models.** These models usually follow the design of the pre-DETR based models [164], [373], which decouples localization and recognition. For object detection, a region proposal network is typically pre-trained for localizing the object of interest [146], [582], and a mask proposal network for extracting masks [137], [515]. Given the localization results, a pre-trained CLIP model is used to measure the similarity between visual contents and language concepts. A clear advantage for two-stage models is that they can inherit the open-set semantic understanding capacity without additional training so as to devote model training to requiring a well-performed localization network.
- **End-to-end models.** Different from two-stage models, the end-to-end models follow the DETR-based methods [43], [81] or other one-stage models [91]. GLIP [242] is one of the representative works. GLIP formulates object detection as textual grounding and is trained end-to-end on image-text pairs with detection and grounding labels. Follow-up works enhance GLIP by enabling deeper vision-language interactions [271] or using DETR-like model design [310], [542]. For segmentation, both ZegFormer [100] and OpenSeeD [557] exploit a DETR-like architecture and predict the masks and categories based on the outputs of their decoders.

Model pre-training. There are mainly three learning methods for pre-training open-set vision models.

- **Supervised learning.** By converting label supervision to language supervision, many works directly leverage the existing supervised annotations for training open-set models. For example, OVR-CNN [543] trains a model with COCO categories and then evaluates its performance on novel categories. Likewise, ViLD [146] trains and evaluates two separate models on COCO and LVIS datasets, respectively. Following a similar protocol, many works train the open-set segmentation models on a subset of annotated segmentation data and evaluate the generalization ability on held-out data [100], [101], [495], [557].
- **Semi-supervised learning.** One might use both annotated data and unlabeled or weakly-labeled data. For example, both RegionCLIP [582] and GLIP [242] use a teacher model to extract fine-grained region-text alignments from image-text pairs to augment the training data for better open-set detection performance. Differently, OpenSeg [138] exploits Localized Narrative datasets [354] as weakly-labeled data, which provides coarse correspondence between language phrases and strokes in images. Empirically, such semi-supervised learning methods often help improve models' generalization ability because they can effectively leverage rich semantics from noisy data.
- **Weakly-supervised learning.** Some works solely use weakly-labeled data for modeling. For example, GroupViT [494] uses a contrastive learning method where all supervisions for model training are from positive and negative image-text pairs. Following the same contrastive learning method, SegCLIP [296] uses a gathering mechanism to learn to merge image patches through the training on image-text pairs.

Below, we review recent models developed for region-level and pixel-level tasks.

4.2.1 Object Detection and Grounding

Object detection is a fundamental task in computer vision that involves identifying and localizing objects of interest within an image or a video sequence [440]. Over the years, various techniques and algorithms have been developed to improve the accuracy and efficiency of object detection. In the past, region-based approaches such as R-CNN [141], Fast R-CNN [140] and Faster R-CNN [373] have been fostering the development of advanced techniques for object detection. To improve real-time performance, YOLO [371] proposes a single neural network that simultaneously predicts object classes and bounding box coordinates. Some improvements are made by either using multiple feature maps at different scales [273] or introducing a focal loss to address the class imbalance problem in dense object detection scenarios [257]. After the emergence of Transformer [437], DETR [43] applies the transformer architecture to object detection, treating it as a set prediction problem. Since DETR, a number of methods have been proposed to improve transformer-based detection models from various aspects, such as DINO [556], Group DETR [61], and Co-DETR [598].

Open-set object detection models aim to detect arbitrary concepts beyond the vocabulary provided in training data. Three main evaluation settings have been developed in the literature:

- **Zero-shot object detection.** Similar to zero-shot image classification [485], zero-shot object detection restricts the object classes used for training, and evaluates models' transferrability to novel classes. Methods falling in this category mainly focus on evaluating how a model leverages pre-trained concept embeddings (*e.g.*, word2vec [309]) and learns good visual-semantic alignments [24], [364], [594], [595].

- **Strict open-vocabulary object detection.** First introduced in OV-RCNN [543], this setting differs from zero-shot object detection in that there is no limit on the training vocabulary as long as it does not cover any target classes. Under this protocol, some representative works are ViLD [146], RegionCLIP [581] which leverage large-scale language-image models [195], [360], and Detic [590] that learns from image-label data.

- **Generalized open-vocabulary object detection.** Some recent works like GLIP [242], and OWL-VIT [310] advocate a more flexible setting to evaluate the dataset or task transferrability for object detection models. This setting allows vocabulary overlap between training and test sets, *e.g.*, Objects365 for training while COCO for evaluation. This is arguably a more practical setting than the two settings described above in that models can be trained using any arbitrary set of training data and their detection performance evaluated in the wild [227].

Object grounding can be considered as a generalized open-set object detection task [56], [93], [202], [353]. In this task, models take a sentence and an image as input and localize objects that are associated with the noun phrases. Recently, M-DETR [198] employs a transformer-based architecture to build an end-to-end modulated detector to detect objects in an image given a raw text query. Unlike previous works where models are trained on specific datasets, the network is pre-trained with 1.3M pairs of text and images, sourced from multi-modal datasets where the connections between text phrases and corresponding image objects are labeled. Inspired by M-DETR, GLIP [242] casts object detection as a grounding problem, and jointly learns a model using object detection and grounding data for open-set scenarios. Following this line of research, DetCLIPv2 [514] proposes a simple joint learning method where multiple tasks are converted into a word-region alignment task, and then a model is trained end-to-end on a corpus consisting of object detection data, grounding data and image-text pairs. Figure 4.4 presents the structures of DetCLIPv2 and Grounding-DINO. Grounding-DINO [271] is a state-of-the-art grounded object detection method, where the object detector is composed of components: a backbone, a neck, and a head, and inject language conditions at every stage. A combined text and image backbone is employed to extract features at multiple scales, which are then passed on to the neck. The text and image characteristics generated by the neck are subsequently used for language-driven query selection. Grounding-SAM is developed by combining Grounding-DINO with SAM [206]. As shown in Figure 4.5, an image and a group of concepts are first fed into Grounding-DINO to produce the boxes, and then the boxes are used as prompts for SAM to predict masks for each box.

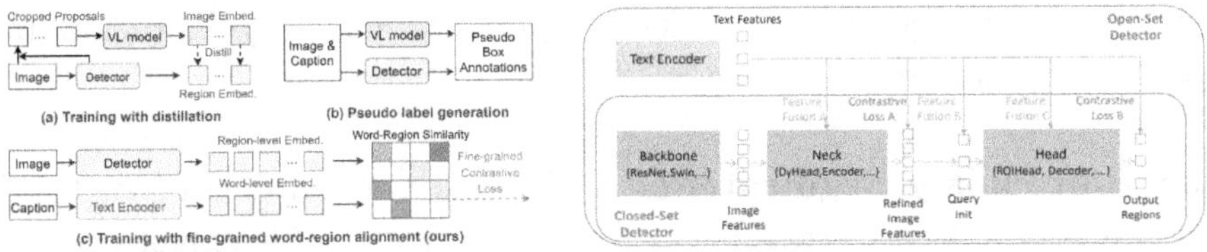

(a) DetCLIPv2 **(b)** Grounding-DINO

Figure 4.4: (a) DetCLIPv2 [514] learns fine-grained word-region alignment from object detection and grounding data and large-scale image-text pairs. (b) Grounding-DINO [271] injects text conditions into different stages of the Transformer encoder-decoder, which significantly improves the grounding performance. Image credit: [514] and [271].

Figure 4.5: Grounding-SAM consists of Grounding-DINO [271] and SAM [206]. Image credit: [271].

4.2.2 Image Segmentation and Referring

Image segmentation is a long-standing and challenging vision problem. There are mainly three subtasks, including semantic [284], instance [160], and panoptic [205] segmentation. Semantic segmentation cares about the per-pixel semantic within an image [57], [80], [284], whereas instance segmentation groups pixels of the same semantic meaning into objects. Models for both tasks have evolved from CNN-based architectures [284] to transformer-based ones [80], and from two-stage models [164] and one-stage models [34], [430] to the recent query-based approaches [102], [600]. With the capability of per-pixel and instance-level understanding, a natural step was taken to formulate panoptic segmentation [81], [205], [445]. Most recently, Mask2Former [81] proposed to address all three tasks with a unified encoder-decoder architecture. Nevertheless, all these works cope with a limited number of categories. In the following, we will review the most recent works on open-set image segmentation and referring segmentation.

Open-Vocabulary Segmentation. Recently, a number of methods have been proposed to transfer or distill the rich visual-semantic knowledge from foundation models [195], [360] to specific segmentation tasks. Prominent examples include LSeg [226], OpenSeg [137], and [189]. Instead of using existing models, GroupViT [494] performs language-image pre-training from scratch with a bottom-up grouping ViT [105], while DenseCLIP [369] demonstrates the superiority of foundation models in finetuning settings compared with supervised models. Recently, MaskCLIP [101] is proposed to tackle open-vocabulary panoptic and semantic segmentation simultaneously by leveraging CLIP, and achieves impressive performance on ADE20K [583] and PASCAL [113], [319]. Instead of using the ViT backbone, a recent work called FC-CLIP [535] exploits a convolutional CLIP backbone (*i.e.*, ConvNeXt trained by OpenCLIP [190]) as both a feature extractor and a vision encoder. Based on a simplified pipeline, FC-CLIP shows plausible efficiency and lefts the state of the art on various open-vocabulary segmentation benchmarks. Rather than only using CLIP, a recent work ODISE [495] leverages text-to-image diffusion models, and shows that the latent features in the pre-trained UNet can provide useful compact segmentation information for open-vocabulary segmentation.

A big challenge in open-vocabulary segmentation is the lack of segmentation data annotated with semantic labels. Thus far, most of the works are still using COCO segmentation annotations. A few recent works attempt to leverage object detection data as the extra supervision to augment the training of segmentation models, such as OpenSeeD [557] (shown in Figure 4.6) and DataSeg [145]. In addition to these new modeling techniques, new datasets have been developed to mitigate this problem, including curating multi-domain segmentation datasets [217], collecting high-quality annotations [292] or scaling up to billions of masks [206].

Referring Segmentation by design is open-vocabulary. Models are usually designed specifically to learn from target datasets using various multimodal fusion strategies [177], [260], [302], [480], [520], [532].

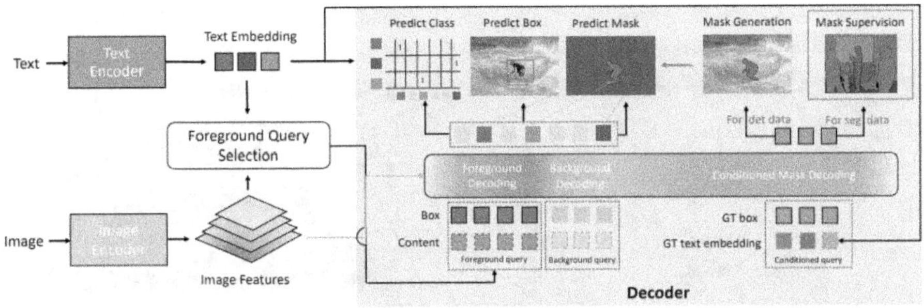

Figure 4.6: OpenSeeD [557] leverages both mask and box supervision for learning a universal open-vocabulary image segmentation model. Image credit: [557].

CLIPSeg [294] extends a textual query to a visual query and shows superior performance not only on referring segmentation but also on semantic segmentation. Since the emergence of vision transformers, works like LAVT [507] enhance the cross-modal interactions from the very beginning, which leads to a decent performance on RefCOCO [532], RefCOCO+ [532] and G-Ref [301], [325]. Differently, PolyFormer [266] converts masks into polygons and asks the transformer decoder to decode a sequence of polygon coordinates. Inspired by Pix2Seq [64], a similar method in object detection, PolyFormer presents an alternative way to represent masks for state-of-the-art referring segmentation. As we discussed earlier, one can also compose Grounding DINO [271] with SAM [206] for referring segmentation.

Unified Segmentation. Given the above methods for open-vocabulary and referring segmentation, an open question is how to unify all segmentation tasks in a single framework. Recently, X-Decoder [599] uses a generalized encoder-decoder architecture to unify all these segmentation tasks. The referring segmentation task is reformulated as a conditioned panoptic segmentation that takes some textual phrases as input to the decoder. UNINEXT [501] is another work that attempts to unify all instance-level segmentation in images and videos. Different from X-Decoder, UNINEXT uses early fusion to fuse the various prompts and vision features, which are then fed to the transformer encoder-decoder.

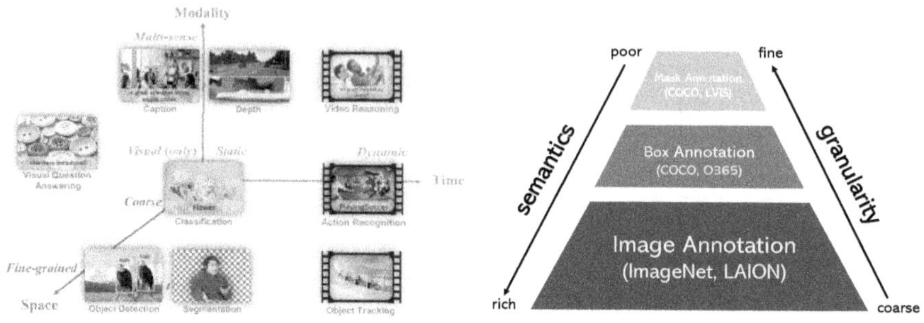

(a) (b)

Figure 4.7: (a) CV task landscape: CV tasks can span different axes, including modality, space and time, which renders significant challenges to unify all of them in a single model. Image credit: [539]. (b) The data scale pyramid: In particular, datasets in different tasks usually contain different types of supervision. Image-level datasets like ImageNet [94] and LAION [386] have annotations that have rich semantics coverage but are coarse-grained, while pixel-level datasets like COCO panoptic segmentation [76] provides fine-grained annotations but with limited concepts.

4.3 From Task-Specific Models to Generic Models

Above we have discussed the recent efforts of transforming closed-set models to open-set ones for detection and segmentation. Until recently, however, most vision tasks have been separately tackled with specialized model designs, preventing the synergy of tasks across different granularities or domains from being exploited. This is arguably due to two reasons:

- **Vision tasks are fragmented.** As shown in Figure 4.7 (a), computer vision tasks span across different axes including space, time, and modality. From the space aspect, it can be image-level, region-level and pixel-level tasks as we discussed before. Along the time axis, we need to tackle not only static images but also temporal video sequences. Regarding the modality, the inputs and outputs can be images, texts, or other types (*e.g.*, human pose, depth map). Such diverse task formats significantly impede the development of a unified model for all tasks.
- **Data scales are different.** In addition to the complicated task landscape, the scarcity of human annotations and their different scales

for different tasks also make building a unified model challenging. In Figure 4.7 (b), we can see a clear pyramid of data scale, where different layers of human annotations have different semantics. More specifically, image-text datasets like LAION [386] contain up to 2B samples, while object detection datasets like Objects365 [389] have 1.7M images in total. More significant gap is observed in segmentation datasets due to the high cost of annotating masks.

Despite the aforementioned challenges, we are now witnessing a growing interest in building unified, general-purpose models that can learn from and be applied to a diverse set of vision and vision-language tasks, thanks to the versatility of transformers [437]. These attempts can be grouped into two main categories:

- **I/O Unification.** Following the development of unified LLMs, a number of recent works reformulate many vision tasks as a sequence-to-sequence problem [65], [287], [451], [509]. They typically use a tokenizer to tokenize the original inputs and outputs (I/O) in different modalities used in various tasks into a coherent sequence (visual or text) tokens and then exploit a unified, sequence-to-sequence model.
- **Functionality Unification.** In addition to I/O unification, one might built a generic model via functionality unification. Extending multi-task learning methods [151], [179], [288], many recent use a coherent encoder-decoder architectures [530], [559], [599]. This line of work usually does not need task-specific or modality-specific tokenizers but requires a sophisticated model design to accommodate various tasks.

Figure 4.8 illustrates the difference between the two categories of unification methods. For I/O unification, the I/O unification module always generates a sequence of tokens, and exploits a separate decoder to decode the final outputs for different tasks. For functionality unification, the functional unification module generates heterogeneous outputs for different task, *e.g.*, semantic outputs and spatial outputs. Then, these different types of outputs are combined to produce the final task-specific outputs. Both unification methods strive to make use of synergy across

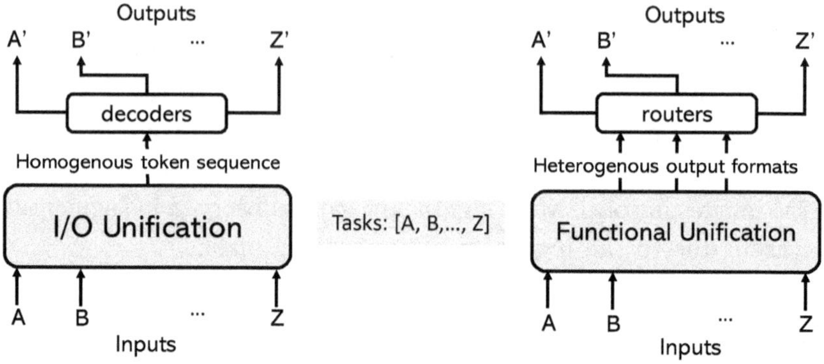

Figure 4.8: Side-by-side comparison between I/O unification and functionality unification. I/O unification is aimed at utilizing a decoder to decode homogeneous token sequences, which are then decoded by task-specific decoders. In contrast, functionality unification predicts heterogeneous outputs and then uses different routers or headers to produce the final outputs for various tasks.

tasks with different levels of granularity. For example, coarse-grained data is expected to contribute to rich semantic understanding required by fine-grained tasks, while fine-trained data to enhance the grounding ability for coarse-grained tasks. In the following, we review some recent works of these two categories.

4.3.1 I/O Unification

This line of work is mainly inspired by LLMs that unify many NLP tasks as sequential modeling. In the vision domain. the methods of building generic models via I/O unification can be grouped into two categories depending on the tasks of interest and output formats.

Sparse and discrete outputs

For vision tasks that produce sparse or discrete token outputs, we can easily exploit a language tokenizer, such as byte-pair encoding (BPE) [388], for I/O unification. In contrast, spatial outputs like boxes, masks, or human skeletons can be formulated as a sequence of numeric coordinates which are then tokenized into discrete tokens [83], [266], [509]. As a result, the decoded output tokens are interleaved with organic

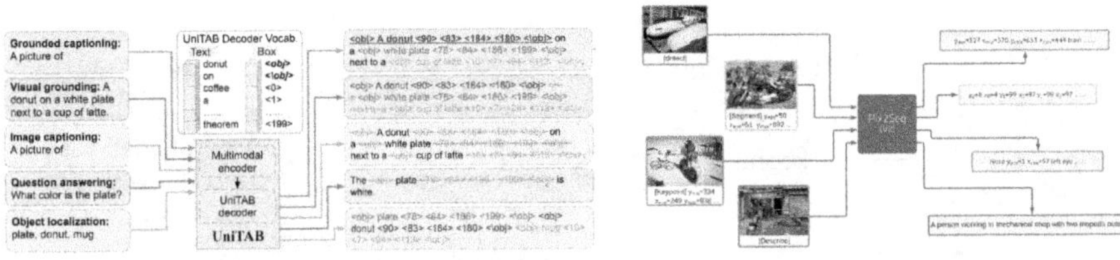

(a) UniTab (b) Pix2Seqv2

Figure 4.9: (a) UniTab [509] is proposed to unify grounded captioning, visual grounding, image captioning, VQA, and object localization. (b) Pix2Seqv2 [65] is proposed to unify object detection, referring segmentation, keypoint detection, and image captioning. Image credit: [509] and [65].

textual tokens and numeric textual tokens to support a wide range of tasks. Without the loss of generality, the decoding process is formulated as auto-regressive generation and the model trained with the objective function defined as:

$$L(\theta) = -\sum_{t=1}^{T} \log p(s_t | s_{<t}, v; \theta), \tag{4.3}$$

where $\{s\}_{t=1}^{T}$ is the discrete token sequence of length T, and v is the visual feature. Below, we review some representative works.

UniTab [509] unifies text and box output in a sequence decoding manner. As shown in Figure 4.9 (a), the box coordinates are represented by numerical numbers with <> and then a special token <obj> is used to encompass the location information. In this way, the model can unify a variety of tasks that require textual and location outputs, including image captioning [76], grounded captioning [353], visual grounding, object localization and visual question answering [8]. The model is trained in three stages: pre-training, multi-task finetuning, and task-specific finetuning.

Pix2SeqV2 [65] slightly differs from UniTab in that it unifies two different vision tasks: referring segmentation and keypoint detection. Following Pix2Seq [64], Pix2SeqV2 represents objects in an image as $[y_{min}, x_{min}, y_{max}, x_{max}, text]$. Then, it introduces a unique task prompt for each task, which contains task type information or a combination of task types and specific locations. For mask decoding, a mask contour is converted into a polygon and then its coordinates extracted from the

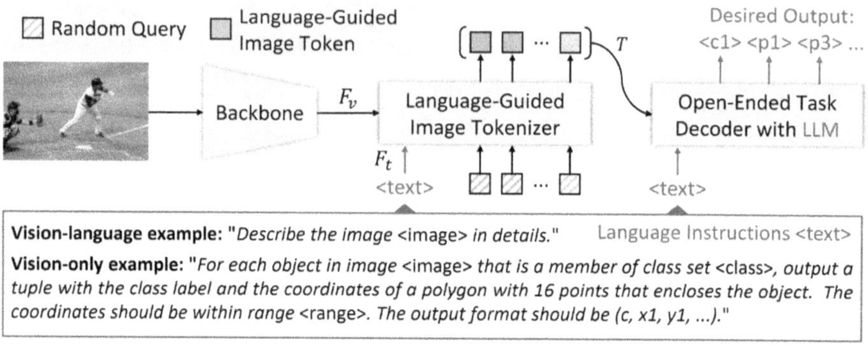

Figure 4.10: VisionLLM [456] is proposed to bridge vision systems with LLMs in a sequential decoding manner. Image credit: [456].

polygon [47]. A similar strategy is also used for referring segmentation, as in Polyformer [266].

LLM-augmented. Recent works have also explored building a generic decoding interface based on LLMs, which are pre-trained on large amounts of text data and human instructions. Kosmos-2 [348] exploits the pretrained LLMs of Kosmos-1 [183] and augments the grounded multi-modal data by collecting a web-scale grounded image-text pair dataset (GRIT) consisting of 91M images. VisionLLM [456] appends an even larger LLM (*e.g.*, LLaMa [433]) on top of an image tokenizer, as shown in Figure 4.10. The resultant model exhibits a very strong vision-language reasoning capacity and decent localization ability for object detection, segmentation, *etc.* Some other works that combine LLMs with grounding are DetGPT [351] and GPT4ROI [566]. To further equip the model with the segmentation capability, both BubaGPT [578] and LISA [215] use an extra referring segmentation model to segment images by taking texts or embeddings as input, respectively. PaLI-X [70] is by far the largest unified model that can cope with multilingual vision and vision-language tasks.

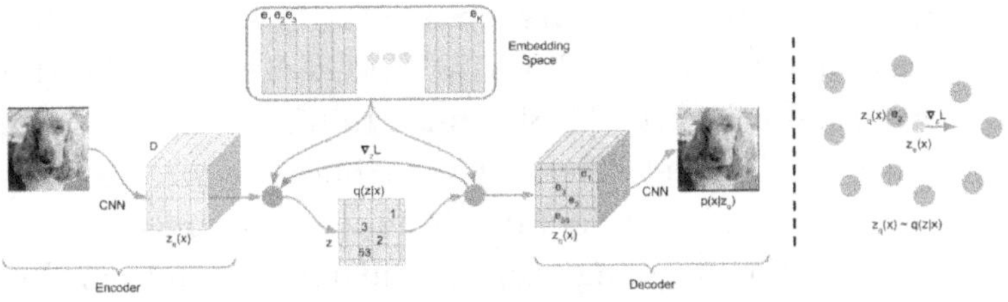

(a) VQ-VAE

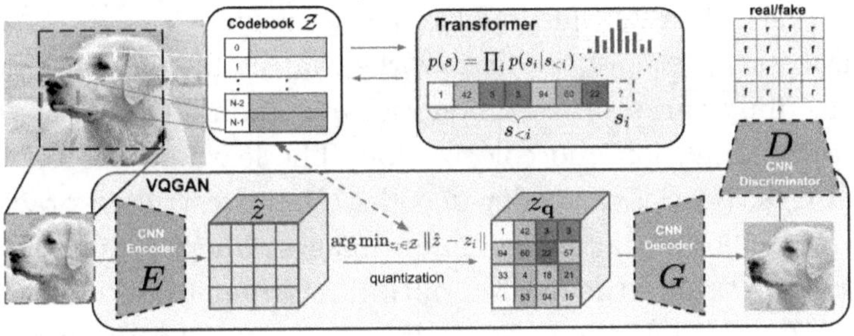

(b) VQ-GAN

Figure 4.11: Illustration of VQ-VAE [332] and VQ-GAN [112].

Dense and continuous outputs

There are also some tasks that require dense and continuous outputs, such as image segmentation [164], depth estimation [307], image inpainting and editing [36], [110]. Except for segmentation masks which can be approximated by polygons [65], [266], most dense and continuous outputs cannot be easily converted into discrete tokens due to the high-dimensional space. Thus, we have to resort to an image-oriented tokenizer. Akin to the language tokenizer, an image tokenizer encodes raw images and extracts discrete tokens spanning the visual feature space. The most representative work is VQ-VAE [332], [370]. As shown in Figure 4.11 (a), VQ-VAE learns an encoder z_e, a decoder z_q and a discrete codebook $\mathbf{e} = \{e_1, ..., e_K\}$ consisting of K embeddings. Given the input x, the posterior categorical probability $q(z|x)$ is defined as:

$$q(z = k|x) = \begin{cases} 1, & \text{for } k = \text{argmin}_i \, ||z_e(x) - e_i||. \\ 0, & \text{otherwise.} \end{cases} \tag{4.4}$$

where the decoder z_q takes x (or its representation e_k) as input to predict class label. As a variant of VQ-VAE, VQ-GAN uses a discriminator and the perceptual loss [216], [218] to maintain a good balance between output quality and model efficiency (via high compression rate). In Figure 4.11 (b), we see that the discriminator is applied at the patch level to regularize the decoding of images at high resolution. Below, we discuss some most recent works that attempt to unify different vision and multi-modal tasks that involve dense outputs.

UViM [211] is one of the first works that employ a dense decoding process to unify various core vision tasks, including panoptic segmentation, depth estimation and colorization. The learning process consists of two stages: (*i*) Base encoder-decoder f and restricted oracle Ω are learned to predict outputs given input images, where f takes raw image as input and Ω takes the desired output as input to decode the oracle code; (*ii*) Instead of using the desired output as input to the oracle Ω, the model learns a language model to produce the oracle code for the input raw image. Notably, the encoder-decoder model used here is trained with VQ-VAE objectives. As the first step to unify vision tasks with a single model, UViM shows promising results on three vision tasks.

Unified-IO [287] is another representative work. Compared to UVIM, it scales to many more vision tasks and datasets. Unlike the training procedure of UViM, Unified-IO first trains different VQ-VAE models for different tasks, as depicted in Figure 4.12 left. After obtaining all VQ-VAE encoder-decoders, 90 datasets are combined to train another transformer encoder-decoder end-to-end, as shown on the right side. Similar to previous works, it also uses a language decoder to obtain the organic and numeric texts to generate coordinate outputs. After the second-stage pre-training, the model achieves state of the art on the GRIT benchmark [156] and exhibits compelling compositionality, although the performance still lags behind the strongest models on common tasks. As a follow-up, a soft-token strategy is proposed in [328] to improve the accuracy for next token decoding. In addition, a masked modeling strategy is proposed to learn robust representations. Evaluated on instance segmentation and depth estimation, the model achieves state-of-the-art performance on NYUv2 [402] and competitive performance

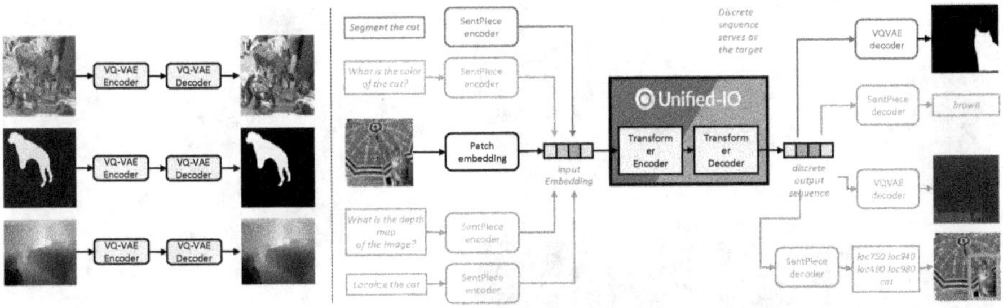

Figure 4.12: Unified-IO [287] unifies different vision tasks by first pre-training VQ-VAE for each task and then an encoder-decoder for tasks jointly. Image credit: [287].

on segmentation. A recent work uses image inpainting as the general task to unify different pixel-level vision tasks [27]. Given the target discrete tokens produced by VQ-GAN, the method exploits a masked autoencoder to decode the missed image regions, using the task input-output examples as prompts. Painter [459] extends this pipeline to facilitate more vision tasks and obtains competitive performance on various standard benchmarks.

Diffusion-augmented. Unlike the above works that learn their own decoding models, some recent works utilize the off-the-shelf stable diffusion model to build generalist vision models. For example, Prompt Diffusion [465] initializes a model using Stable Diffusion and ControlNet [562], and trains the in-context image-to-image model jointly on six different vision-language tasks, including segmentation, depth estimation, *etc.* InstructDiffusion [136] also uses the diffusion model but explicitly introduces task-specific instructions to the diffusion process. Moreover, it uses task-specific training and human alignment training to enable a generalist interface for vision tasks.

4.3.2 Functionality Unification

Unlike I/O unification, functionality unification attempts to unify different tasks based on the task characteristics, with the awareness that they are neither fully isolated nor fully aligned. At a high level, vision tasks produce three types of outputs: (*i*) location outputs, (*ii*) semantic

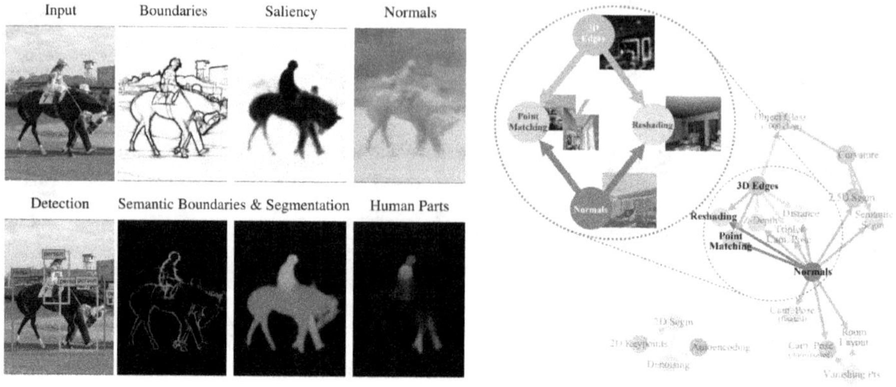

(a) UberNet (b) Taskonomy

Figure 4.13: (a) UberNet [209] can be applied to 7 vision tasks using a unified and budget-controllable CNN architecture. (b) Taskonomy [540] further studies the relationship across vision tasks by exploiting a multi-task transfer modeling. Image credit: [209] and [540].

outputs, and (*iii*) pixel-level outputs. For example, both object detection and phrase grounding need to localize objects in the image, while both generic segmentation and referring segmentation produce masks. On the other hand, many tasks require semantic (or text) outputs to represent either concept names or textual descriptions.

Multi-task learning

Some early works explore multi-task learning methods for unifying different vision or vision-language tasks.

Vision models. A few works explore using CNNs for learning with different vision tasks at different levels. For example, Cross-stitch Networks [312] develops a strategy to split different numbers of layers from the top in CNNs so as to adapt to different vision tasks. Results show that the best-performing multi-task architecture depends on the tasks of interest and can hardly generalize to new tasks. UberNet [209] takes one step further to use a single universal CNN architecture and sophisticatedly design a routing mechanism to save the memory and computing cost, as shown in Figure 4.13 (a). Both works require some tweaking to the CNN architecture so that they can adapt to different levels of tasks and loss types. But they unfortunately fail to build the synergy across tasks to improve model performance. Taskonomy [540] specifically studies the relationship among vision tasks. It first trains task-specific models for each individual task and then performs transfer modeling across tasks in the latent space. The task affinity is then calculated in the latent space, providing us with the taskonomy. The result shows that vision tasks have different affinities for different groups, as shown in Figure 4.13 (b). For example, surface normal estimation is heavily related to reshaping and point matching. Curvature extraction is related to image segmentation tasks. This study provides deep insights for multi-task vision modeling [87], [492].

Multi-modal models. The emergence of Transformers significantly facilitates the advancement of multi-task multi-modal learning. Among them, 12in1 [288] is one of the pioneering works that combine 12 vision-language tasks in a single BERT-based architecture. It uses task-specific heads for individual tasks and a commonly shared trunk ViLBERT [286]. Results show that multi-task learning can achieve substantial improvements over single-task learning while reducing the model parameters significantly. Later on, UniT [178] exploits an encoder-decoder architecture and expands to vision-only tasks like object detection. Additionally, it allows end-to-end training on the task pool without relying on pre-trained detectors. Similar to 12in1, it also uses a task-specific head for each task, motivated by the empirical result that sharing the same head usually hurts performance. Likewise, E2E-VLP [493] proposes an end-to-end pipeline for both localization tasks and text generation. Both UniT and E2E-VLP demonstrate the versatility of the encoder-decoder architecture of DETR [43]. Following the same spirit, GPV [152] proposes an end-to-end task-agnostic architecture for different vision and vision-language tasks. It uses DETR to extract boxes and region features and then exploits a cross-attention module for fusion, followed by a vision decoder and a language decoder for decoding different outputs.

The above vision and multi-modal models unify different tasks by incorporating different modules or heads designed to cope with different tasks, and can hardly achieve synergy across tasks. In the following, we discuss recent model unification research that aims to make the best use of synergy among various vision and multi-modal tasks.

Unified learning

The barrier across tasks is gradually blurred thanks to the use of Transformers [437] and the development of open-set models as we discussed earlier. It is now possible to bind inputs from different modalities to learn a shared semantic space. A number of works [233], [559], [599] have recently been proposed to unify vision and vision-language tasks by *using one model for all*. After pre-training, the single model can be applied to tackle all tasks in a zero-shot manner and the performance can be further improved via task-specific finetuning. Note that unified learning in this context differs from previous works of large-scale pre-training. Like GPT which serves as a universal language interface after pre-training, a unified vision model is not only a representation learning engine but also an interface that supports as many tasks as possible in a zero-shot manner. Below, we review a few representative works.

GLIPv2 [559] extends GLIP [242] to support a wider range of vision and vision-language tasks, including grounded captioning, visual question asnwering, *etc.* GLIPv2 seamlessly integrates localization pre-training and Vision-Language Pre-training (VLP) through three distinct pre-training tasks: (*i*) phrase grounding, which serves as a vision-language adaptation of detection tasks; (*ii*) region-word contrastive learning, where a novel contrastive task at the region-word level is used; and (*iii*) masked language modeling. In a zero-shot manner, this pre-trained model can be applied to different tasks and attain plausible performance across the board. Unlike previous works (*e.g.*, GPV [152]), it merges the localization module and the vision-language matching module in a coherent manner, significantly improving the efficiency and effectiveness of model training using fused data.

X-Decoder [599] follows the generic design of encoder-decoder architecture. Given an input image, it first uses an image encoder to extract features at multiple scales. Afterward, a text encoder is used to encode a textual query into a sequence of embeddings. The visual features, textual queries and the non-semantic or latent queries are fed to a decoder to predict the outputs. Three critical designs are proposed to allow X-Decoder to easily generalize to a variety of vision and vision-language tasks: (*i*) It defines two types of queries and outputs. Specifically, the queries for the decoder are categorized into latent queries and text queries, which undertake generic vision and vision-language tasks, respectively. Likewise, the output is categorized into pixel-level masks and semantic embeddings; (*ii*) A single text encoder is exploited to encode textual corpora for all tasks. The common text encoder is used to encode referring phrases, text descriptions, and image captions for the tasks of referring segmentation, image-text retrieval and image captioning, respectively; (*iii*) It fully decouples the image and text encoders, and use all the outputs as queries. As such, it can learn from both intra-image and inter-image supervisions. This is essential to learn strong pixel-level representations and support different granularities of tasks. As shown in Figure 4.14, the pre-trained model supports different tasks by taking different routing but sharing the same suite of parameters.

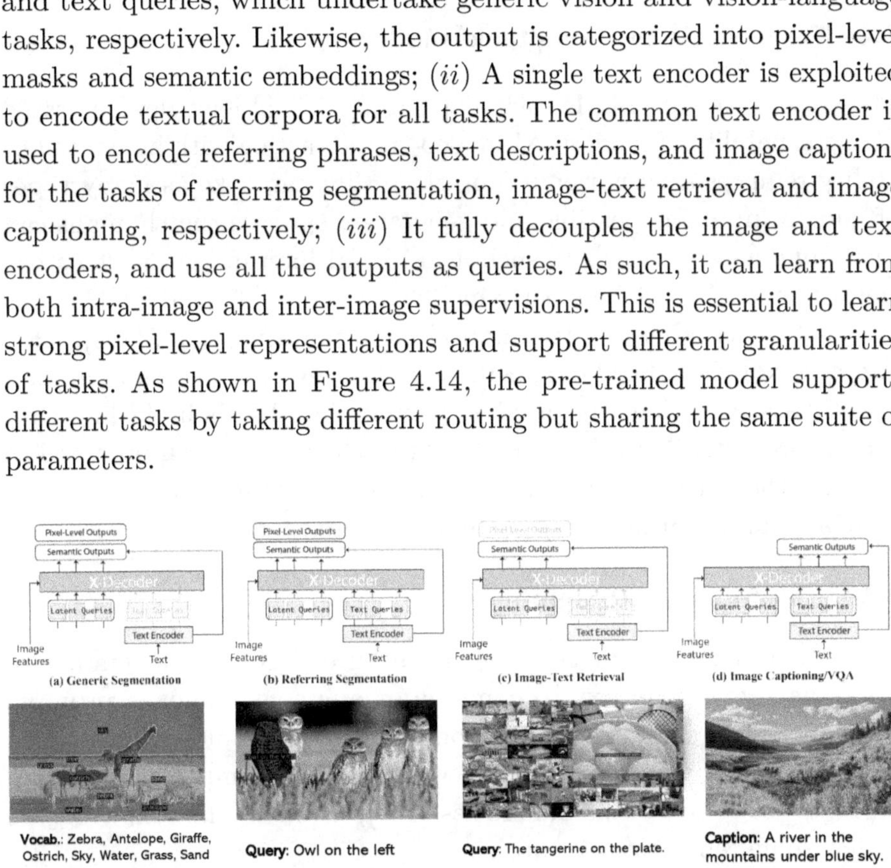

Figure 4.14: A generalist model X-Decoder [599] that unifies different vision and vision-language tasks in a functional manner. It uses a single decoder with the same suite of parameters, but different routing mechanisms to tackle different tasks. Image credit: [599].

Uni-Perceiver-v2 [233] is another generalist model that unifies vision and vision-language tasks. Similar to X-Decoder, the model exploits a vision encoder, a text encoder and a general decoder. Differently, it introduces a region proposal network on top of the vision backbone to explicitly predict the boxes and masks, which are then encoded as "queries" for the general decoder. To jointly train on datasets of different tasks, a unified max-likelihood estimation strategy is used for the tasks with and without localization.

4.4 From Static to Promptable Models

The state-of-the-art LLMs such as ChatGPT [335] have demonstrated strong capability of modern AI models in interacting with humans. The ability requires a user-friendly interface that can take as many types of human inputs as possible and generate responses that humans can easily understand. In NLP, such a universal interaction interface has emerged and evolved from early models like GPT [37] and T5 [363], to more advanced querying techniques like prompting [245], [400], [580] and chain-of-thought [208], [384], [471]. However, most vision models are still static in that they are less flexible than LLMs in dealing with various prompts. Recently, a number of methods have been proposed to enhance the static vision models with the capabilities to support: (*i*) multi-modal prompting; (*ii*) in-context prompting.

4.4.1 Multi-modal Prompting

Vision is different from language by nature. Recently, a number of works have explored how to combine or augment language prompts with other types of prompts to complement incomplete information or resolve language ambiguity, such as spatial prompts [206], visual prompts [601] and other modalities [139], [269]. In the following, we review some representative works.

Spatial prompting. Vision is rooted in the physical world, and as such it is not only semantic but also spatial by nature. Spatial prompting can be considered as a way to modulate the vision models through the inputs of location information,which could be a point, a box, or an arbitrary stroke, *etc.* Such clues have been heavily used in UI designs of computers (*e.g.*, mouse) and mobile devices (*e.g.*, touch screen). In computer vision, interactive segmentation [72], [73], [305], [317] naturally

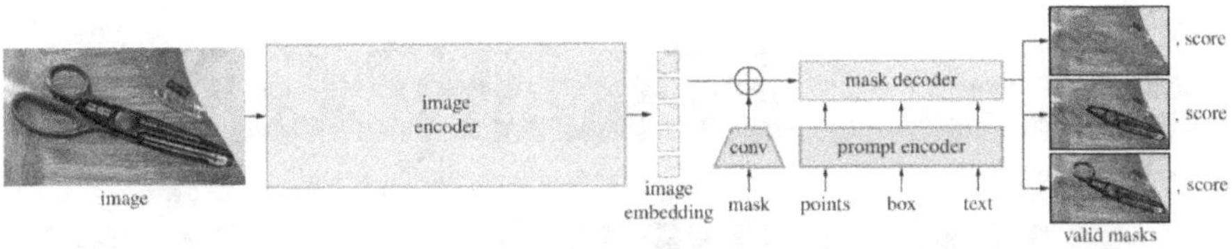

Figure 4.15: SAM [206] proposes a promptable segmentation model which can take different spatial prompts in addition to text prompts. It further develops a data annotation engine to scale up the mask-annotated data. Image credit: [206].

requires such capability so that the model can take multiple user inputs (*e.g.* clicks) to gradually refine the segmentation mask. However, most of these works are still designed task-specifically and lack enough flexibility to support different types of spatial prompts.

SAM [206] is one of the pioneering works that propose a convenient spatial prompting interface and learn a foundation model for image segmentation. As shown in Figure 4.15, the model can take points or boxes as prompts, and segment images at arbitrary levels of granularity. The ability to segment images to follow user instructions makes SAM the foundation model for many vision models and applications [552]. For example, a number of promptable segmentation models have been developed in the medical domain [299], [378] are based on SAM. Spatial prompting is particularly useful in that the textual annotations for medical images are usually limited and hard to interpret. Similar cases also happen in other industry domains [422]. To further improve point prompting, SAMAug [89] refines the points using the max-entropy criterion and saliency map to determine the most informative locations the model should look at.

Visual prompting. In many cases, textual descriptions of objects are not sufficient to convey all important information. Given an unrecognizable or indescribable object, people often fail to express themselves clearly about the object using language. In such cases, expressing by showing examples is more informative and straightforward. With this idea, a line of works have studied exemplar-based visual modeling, such as image-to-image retrieval [92], [527], [561], image co-segmentation [194], [197] and visual object tracking [298], [483], [524]. Recently, this strat-

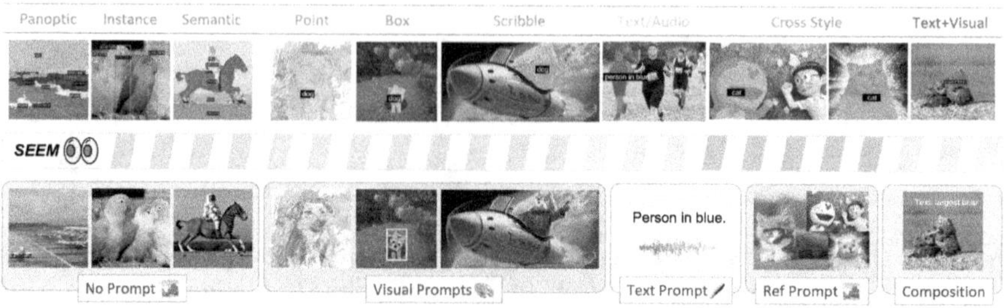

Figure 4.16: SEEM [601] can take different types of prompts as inputs for various image segmentation tasks. Image credit: [601].

egy has been formulated as visual prompting in that different types of visual inputs are encoded using a unified format and then fed into a Transformer model, as shown in LLMs.

SEEM [601] is one of the representative works that enable visual prompting to a vision model for image segmentation. As shown in Figure 4.16, SEEM differs from the aforementioned SAM and can take visual prompts by drawing points, boxes, and strokes on an image that can be either the target image or a reference image. It develops a new module, called *visual sampler*, to extract visual features from an image according to the locations specified by users. Based on the visual sampler, the model can even take another reference image as input without re-training. SEEM shows impressive performance not only for various image segmentation tasks but also for video object segmentation in a zero-shot manner.

PerSAM [564] is a personalized segmentation model, developed based on SAM, which takes one shot as input. A specific model is learned to take a source image and a mask as input, and predicts the mask for a target image. To extract the visual prompts, mask pooling is performed when feeding the input tokens to the decoder of PerSAM. It also extracts the positive and negative priors based on feature matching to provide the pre-trained SAM model with comprehensive clues. Like most prompt learning methods in LLMs, PerSAM can be easily attained by off-the-shelf models like SAM. SAM-PT [365] extends this strategy to video object segmentation. Inspired by spatial prompting of SAM, it exploits a point-tracking system [161] to track different points (both positive and negative) and ask SAM to segment an image given the points. It exhibits strong performance of point tracking and segmentation.

Others. Some other works combine a wide range of visual prompting types. For example, Painter [459] reformulates different vision tasks (*e.g.*, depth estimation, image segmentation) as prompting and learns a decoder in an in-context learning manner. The prompts are combinations of raw images and their corresponding dense annotations (*e.g.*, depth or segmentation maps). In contrast, Prismer [269] makes use of off-the-shelf vision models to extract different information from raw images and then feed the information to a vision-language model. To facilitate the interplay across multiple modalities, ImageBind [139] learns a universal alignment among image/video, language, audio and depth. Then the learned embedding space is used to compose different types of prompts via summation.

4.4.2 In-context Prompting

In-context learning has been observed in many LLMs such as GPT-3 [362]. The ability makes LLMs configurable via prompting without any model parameter updates. In comparison with LLMs, the in-context learning capability for vision models is still less studied. Flamingo [4] is one of the pioneering works. It demonstrates that the ability of in-context language generation for multi-modal inputs can be acquired by model training on interleaved image-text pair data. Likewise, Kosmos-1 [183] treats visual inputs as a foreign language so that the in-context learning ability in LLMs can be naturally transferred to multi-modal inputs. However, both methods take multi-modal data as input but merely generate texts as output. As we discussed earlier, vision tasks require outputs of different types beyond texts. How to endow vision systems with the in-context learning ability for is still an open question. Below, we review some recent attempts toward the goal.

Visual prompting via inpainting is proposed in [27] to teach the model to predict dense outputs, such as edges, masks, depths, *etc.* As shown in Figure 4.17, given an input image $\mathbf{x} \in \mathcal{R}^{H \times W \times 3}$ and a binary mask $\mathbf{m} \in \{0,1\}^{H \times W}$, an inpainting model predicts the missing region $\mathbf{y} = f(\mathbf{x}, \mathbf{m})$. The authors exploit a pre-trained VQ-GAN to encode the original image into discrete tokens and use another ViT encoder to predict the masked regions. To ensure that the model understands the

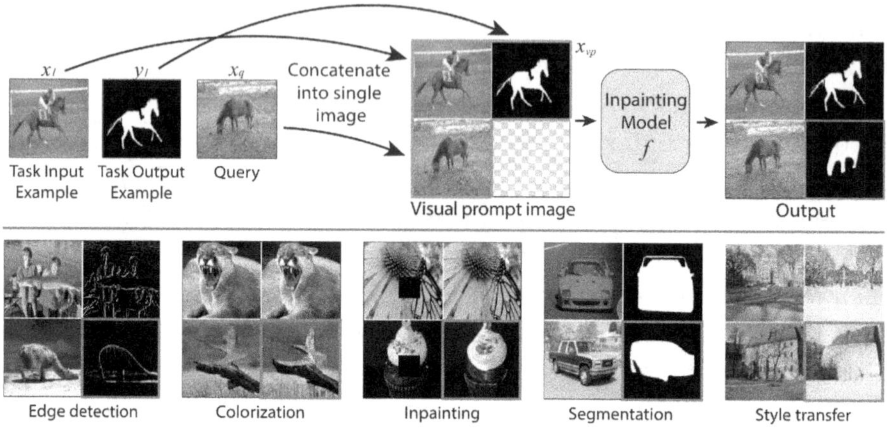

Figure 4.17: Visual prompting via visual inpainting proposed in [27]. Image credit: [27].

visual "context" in the images, the authors collected a new *Computer Vision Figures* dataset which consists of 88k images from Arxiv papers. After pre-training, the model is used to predict the content at the bottom-right corner.

Concurrently, Painter [459] extends visual in-context learning to a diverse set of datasets and benchmarks. Unlike [27], Painter predicts the output in a continuous pixel space instead of discrete token space. For different tasks, different rules are used to convert the output spaces into image spaces. For example, different colors are used to represent different individual instances in the image for the segmentation task. After unifying the input and output format, the vanilla ViT is used as the encoder and masked image modeling [162]. As a follow-up, SegGPT [460] is built on top of Painter and designed specifically for image segmentation tasks. The pre-trained model can be easily extended for exemplar-based image segmentation tasks. Please see Figure 4.18 for the in-context learning task examples of SegGPT.

Hummingbird [23] resorts to a different method for in-context visual learning. Instead of using masked modeling, the authors propose to leverage attention across target and source images to aggregate the information. As shown in Figure 4.19, the model takes multiple input images (first row) and their corresponding semantic label maps (second

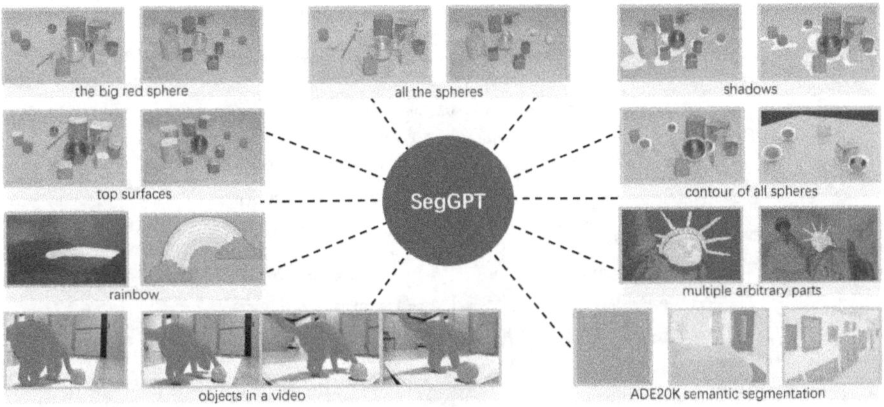

Figure 4.18: SegGPT [460] proposes to perform in-context learning for image segmentation. Image credit: [460].

Figure 4.19: Hummingbird [23] is proposed for in-context visual scene understanding through nearest-neighbor retrieval. Image credit: [23].

row). Given a query image, it first finds the nearest neighbor feature locations in the prompt images for the query points and then projects the same matches to the semantic label maps so as to aggregate the labels for the target query. This strategy is akin to earlier works on building classifiers based on K-nearest-neighbor but differently applied to dense prediction tasks.

Discussion. In-context learning is an appealing feature for AI models. On the one hand, there are a number of works that attempt to extend the in-context learning capability from language models to multi-modal models such as Flamingo [4] and Kosmos-1 [183]. On the other hand,

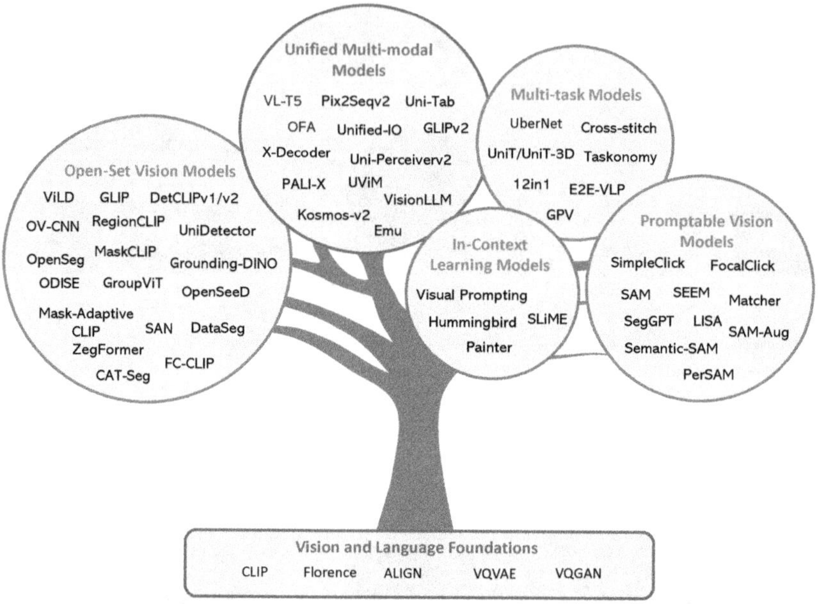

Figure 4.20: A summary of topics covered in this section. A lot of effort has been spent to unify vision models from different aspects to enable more intelligent vision systems.

researchers resort to pure vision-based in-context learning to address vision-specific tasks including image segmentation, depth estimation, *etc.* Thus far, there is no single model that can take multi-modal inputs and predict different types of outputs in an in-context learning manner. This is a promising direction for future research.

4.5 Summary and Discussion

A summary of the works discussed in this section is illustrated in Figure 4.20. There is a clear trend in the vision community to build *open-world, unified, interactive* vision models. Nevertheless, there are still some intrinsic differences between vision and language. First, vision captures the physical world with raw signals while language with man-made symbols. We need to develop sophisticated visual tokenization methods to compress the raw data into compact "tokens". In the language domain, this can be easily done by using some well-established heuristic tokenizers [388]. Second, unlike language, vision data is often not labeled and thus is difficult to decipher its conveyed information or knowledge. It always requires human labors to annotate the visual contents via semantic or spatial labels. Third, language data is homogeneous while vision data and tasks are heterogeneous. Last but not least, storing vision data is much more costly than that of language data. For example, GPT-3 consumes 45 TB of training data, while storing the ImageNet dataset which contains 1.3M images costs more than hundreds of gigabytes. When it comes to video data like Howto100M [308], the storage cost exceeds that of training corpora used for GPT-3. All these differences cast challenges that need to be addressed in the vision community, as detailed below.

- **Computer vision in the wild**. Due to the heterogeneous nature, the current vision data we use for training models can hardly cover the whole physical world. Despite the effort in building open-set vision models, we are still facing significant challenges in coping with novel or long-tail objects and scenarios.
- **Scaling law in vision**. As discussed in [174], [200], the performance of LLMs improves smoothly with the increase of model size, data scale, and compute resource. As the scale increases, some intriguing emergent abilities are observed in LLMs. In contrast, it is still not clear what the right path to scale vision models is, not to mention the emergent abilities in such models.
- **Vision-centric or language-centric models**. Currently, the boundary between vision and language is gradually dismissed. However, due to intrinsic differences between vision and language, it remains open whether we should scale up the vision models as we do for LLMs, or integrate LLMs with moderate vision models to address most (if not all) of the problems.

Despite the progress, we are still far away from an intelligent vision system that can perceive the world like humans. We hope that the literature review in this section provides a comprehensive picture of existing efforts, and inspires the development of next-generation vision models.

5

Large Multimodal Models: Training with LLMs

This section comprehensively reviews the works of building large multimodal models (LMMs) [*e.g.* 4], [334]. Section 5.1 presents the background of multimodal models, focusing on the basics of image-to-text generative models and their representative model instances in various case studies. The state-of-the-art OpenAI Multimodal GPT-4 [334] and the existing research gaps in the field are discussed. To better understand instruction tuning in LLMs, Section 5.2 examines its importance and its role in self-instruct and open-source LLMs. Moving forward, Section 5.3 describes instruction-tuned LMMs, shedding light on their basics, significance and applications. Section 5.4 presents advanced topics in the realm of LMMs. Finally, Section 5.5 discusses the progress in the research community on developing open-source models that perform as good as the proprietary OpenAI Multimodal GPT-4 model. Figure 5.1 illustrate an AI model standing a library full of books, representing a LMM built on a LLM that has learned from a vast amount of language data.

Figure 5.1: An AI model is depicted being standing in a library brimming with books, symbolizing how a LMM is developed from a LLM that has processed extensive language data.

5.1 Background

5.1.1 Image-to-Text Generative Models

LMMs in their current form are primarily image-to-text generative models, which take images as input, and output a text. One example is illustrated in Figure 5.2 (a) Left. All of the model variants share a very similar model architecture and a training objective.

- *Model Architecture.* As illustrated in Figure 5.2 (a) Right, the model typically consists of an image encoder to extract visual features, and a language model to decode the text sequence. The vision and language modalities can be optionally connected by a trainable connection module. The image encoder and language model can be either trained from scratch or initialized from pre-trained models.
- *Training Objective.* As illustrated in Figure 5.2 (b), it typically employs an auto-regressive loss on the output text tokens. For the attention map in the Transformers [437], image tokens can attend to each other, and the current text token attends to all image tokens and the previous text tokens.

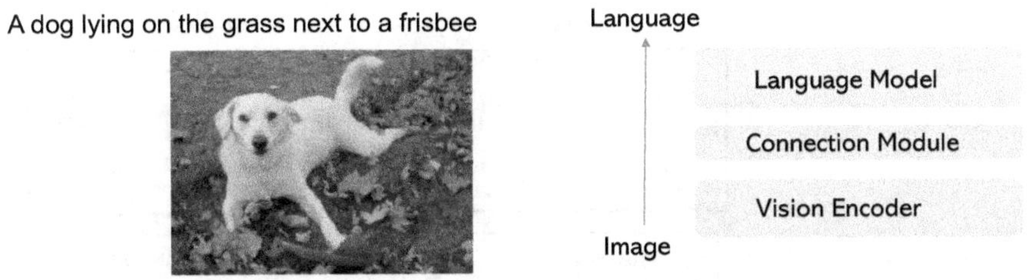

(a) Left: An example of image-to-text generation task; Right: Model architecture.

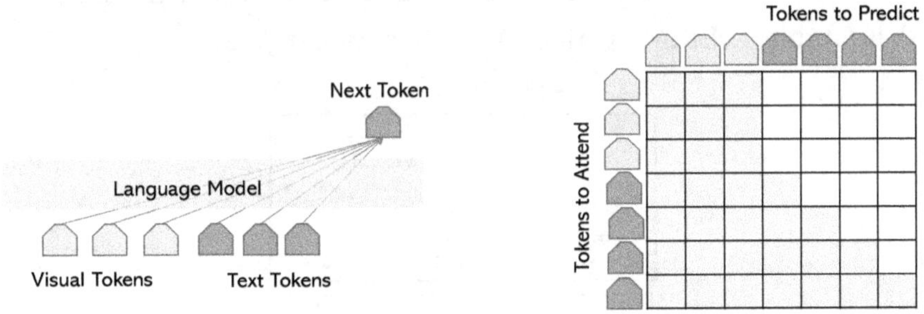

(b) Training objective and attention mask. For each row, the yellow elements indicate that the prediction token attends the tokens on the left.

Figure 5.2: Illustration of image-to-text generation task, architecture, and training objective.

5.1.2 Case Studies

We use some prominent LMMs as examples to illustrate how the aforementioned network architecture can be instantiated in different models, while maintaining the same auto-regressive training objective.

Case study I: LMMs trained with image-text pairs. Most LMMs are trained on a large number of image-text pairs. GIT [446] and BLIP2 [234] are two large models that achieve state-of-the-art (SoTA) performance on many datasets. The comparisons are shown in Figure 5.3(a). GIT initializes its image encoder with the contrastively pre-trained Florence model [539], and trains the language model from scratch. On the other hand, BLIP2 freezes the weights of a pre-trained image encoder and a

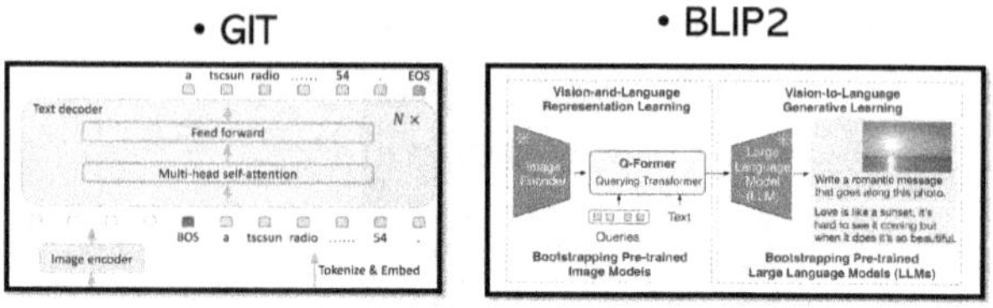

	GIT	BLIP2
Language Model	From Scratch	Pre-trained: FLAN-T5/OPT
Connection Module		Q-Former: Lightweight Querying Transformer
Vision Encoder	Contrastive pre-trained: Florence/CLIP	Contrastive pre-trained: EVA/CLIP

(a) Example 1: LMMs trained with image-text pairs.

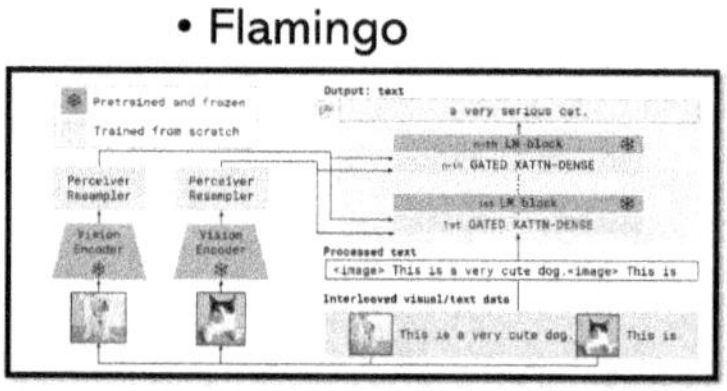

	Flamingo
Language Model	Pre-trained: 70B Chinchilla
Connection Module	Perceiver Resampler Gated Cross-attention + Dense
Vision Encoder	Pre-trained: Normalizer-Free ResNet (NFNet)

(b) Example 2: LMMs trained with image-text pairs and interleaved image-text data.

Figure 5.3: Examples of image-to-text generation models. Image credit: [4], [234], [446].

pre-trained language model, and trains a lightweight Q-former module to connect the image encoder and the language model.

Case study II: LMMs trained with interleaved image-text sequence instances. We use Flamingo [4] as an example, shown in Figure 5.3(b). It connects a frozen pre-trained image encoder and a language model – by adding novel architectural components in between. Specifically, the Perceiver Sampler module helps reduce computational complexity, and the Gated Transformer module helps to stabilize training during the initial stage. Flamingo is trained on large-scale web corpora containing arbitrarily interleaved text and images. After the training is done,

Flamingo can be adapted to downstream vision tasks via in-context few-shot learning (*e.g.* prompting the model with task-specific examples) without additional task-specific tuning.

Multimodal in-context-learning. Beside the SoTA performance on dozens of academic benchmarks, probably the most appealing aspect of Flamingo is the emergent ability: Multimodal In-Context-Learning. Specifically, given a couple of image-text pairs as examples, Flamingo can perform a new task in a few-shot manner, such as solving visual math problems. This means Flamingo can tackle a number of difficult problems with just a handful of task-specific examples, without model re-training. Figure 5.4 shows two new tasks to Flamingo. The top row provides two image-text pairs as the context in the prompt, where the text describes the name of the animal in the image, followed by the geographical information of the animal. Flamingo is able to understand the patterns presented in the examples, and output the description for a new image. In the bottom row, the text first shows the optical character recognition (OCR) result of the image, followed by the answer to a math problem. Flamingo follows the multi-modal instruction, and generates the correct answer for a new math problem in the third image. This intriguing in-context learning capability makes Flamingo the GPT-3's LMM counterpart.

Figure 5.4: The emerging property of pre-training on web-scale interleaved image-text data: multimodal in-context-learning. Examples are adopted from [4].

5.1.3 OpenAI Multimodal GPT-4 and Research Gaps

In March 2023, OpenAI releases GPT-4 [334], which demonstrates very impressive capabilities in visual understanding and reasoning. Though the model details are not revealed, GPT-4 powers many new scenarios based on the examples highlighted in the technique report. Two popular visual examples are illustrated in Figure 5.5. In the first example, GPT-4 identifies the uncommon visual region and exhibits a strong complex reasoning ability. In the second example, GPT-4 recognizes text in the image and captures the meme across image-text. As the research community has no clue how this new ability is achieved, it naturally raises a question: how the community can work together to build open-source LMMs that work like the proprietary Multimodal GPT-4 model.

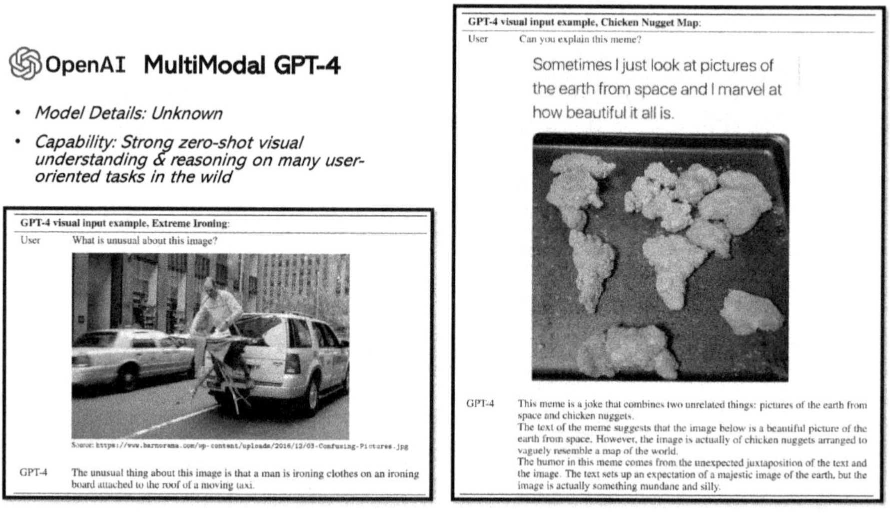

Figure 5.5: OpenAI Multimodal GPT-4. Visual examples are from [334].

Figure 5.6 illustrates some of the most appealing properties of OpenAI models. Our key observations are: (*i*) GPT-2 [362] is the auto-regressive counterpart of BERT [98] which is based on the pre-train-then-finetune paradigm. Compared with GPT-2, GPT-3 [37] is a 175B model trained on web-scale text corpora. GPT-3 exhibits two emergent abilities: in-context-learning [37] and chain-of-thoughts (CoT)

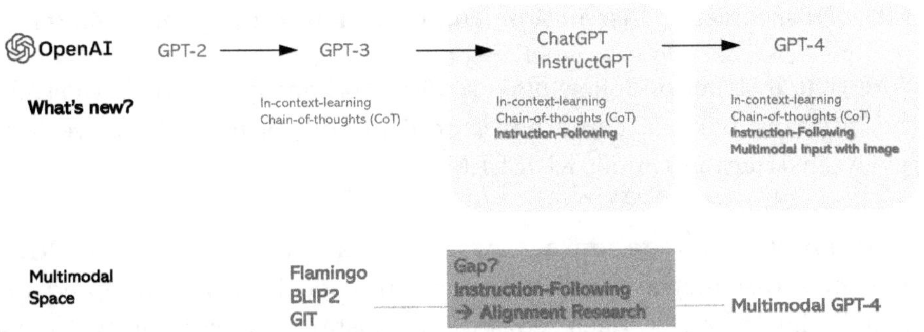

Figure 5.6: Recap on Language Modeling: OpenAI LLM development history. The unique properties for each generation model are highlighted, from which the research gap is revealed for LMM.

reasoning [471]. This means that GPT-3 can tackle a wide range of new problems with just a few task-specific examples or by properly prompting it step-by-step without model re-training, It thus changes the modeling paradigm from task-specific finetuning to prompting frozen models. The latter has higher generalizability and a lower cost of adapting to a new task than the former. (*ii*) ChatGPT and InstructGPT [338] show that it is important to align LLMs' responses with human intents by finetuning GPT-3/GPT-3.5 on high quality instruction-following data and continually improving them using a reward model via reinforcement learning with human feedback. (*iii*) GPT-4 not only improves language understand and generation abilities, but also can take visual signals as input for visual understanding and reasoning.

Figure 5.6 also shows that the newer models always maintain and improve the capabilities of earlier ones, and add new abilities. For example, compared to GPT-3, GPT-4 has two new abilities: instruction-following and multimodal input. This reveals the gap between existing LMMs (*e.g.* Flamingo) and multimodal GPT-4. How to align LMMs with human intents via instruction tuning in the multimodal space is the main topic to be discussed in this section.

5.2 Pre-requisite: Instruction Tuning in Large Language Models

Note that instruction-following is a notion originated in NLP. To study the intuition behind it and have a full picture of its history, we first review instruction tuning with LLMs.

Traditional language data. As a typical data instance in NLP, sequence-to-sequence (seq2seq) representation is widely adopted for many language tasks. Each seq2seq data instance consists of two parts: one sequence as input and the other as output. Figure 5.7 (a) shows two seq2seq examples of translation and summarization, respectively.

Seq2seq data is fundamental to building many NLP models. One can either build individual models, one for each NLP task, or build one model for multiple (related) tasks via multi-task learning. Either way, the trained NLP models are hard to generalize to new tasks in a zero-shot fashion, as they are not trained to understand task instructions and thus cannot be applied to new tasks on the fly.

Instructional language data. Recently, researchers have started to explicitly use task instruction data for model training, as shown in Figure 5.7 (b). Interestingly, the task instructions for a majority of NLP tasks can be expressed in natural language. It leads a new training data format: instruction-input-output triplets. Based on the new format, one single model can be trained to perform multiple tasks, each with its task-specific instructions. Since models have observed task instructions and many instances for each task during training, it is possible for them to generalize to new tasks by task composition in the inference stage.

Figure 5.7 (c) shows a compositional task that requires both summarization and translation. Though the model has never seen this compositional task during training, it is trained on data from the two individual tasks. Thus, the model can perform the new task by composing translation and summarization. Note that humans are always creating new tasks in our daily life, which are unlikely to be observed by models during training. It is thus appealing to have an AI model that can deal with thousands of new tasks in the wild without re-training. This is the main reason why ChatGPT becomes so popular and prevalent.

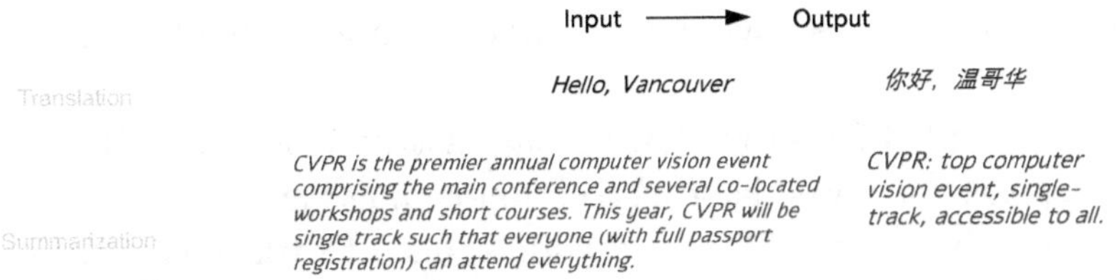

(a) Training: Implicit task instructions in traditional language data.

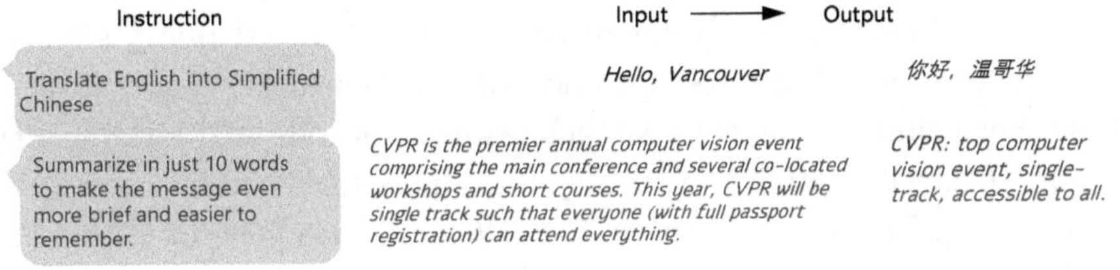

(b) Training: Explicit task instructions in instructional language data.

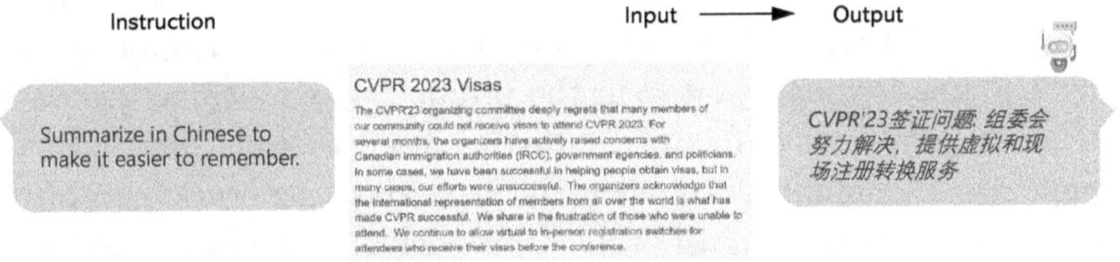

(c) Inference: Explicit task instructions in instructional language data. The English meaning of the output is *"CVPR'23 visa issue: organizing committee works to solve and provide virtual and in-person registration switch services."*

Figure 5.7: Examples of task instructions in traditional and instructional language data, respectively.

5.2.1 Instruction Tuning

We can collect a diverse set of high-quality instruction-following data using two methods. One is through human-human interactions, where humans (task providers) provide an annotation statement and requirements, based on which another group of human subjects complete the annotation tasks. This method is typically costly and time consuming. The other method is to collect data via human-machine interactions, where task providers provide the annotation statement and requirements, but it is AI systems or models that complete the annotation tasks.

To enable LLMs to follow human instructions to complete real-world tasks, various methods of instruction-tuning have been explored. These methods finetune LLMs on a wide range of tasks either using human-annotated prompts and feedback [338] or using public benchmarks and datasets augmented with manually or automatically generated instructions [463]. Instruction-tuning research has produced many effective methods for improving zero-shot and few-shot generalization abilities of LLMs. For example, the self-instruct tuning method [462] leverages the in-context-learning ability of LLMs, and aligns LLMs to human intents by finetuning LLMs using instruction-following data generated by SoTA LLMs. The method is illustrated in Figure 5.8. Humans create a few examples (*i.e.* seed examples) as the context, and LLMs such as GPT-3 or GPT-4 are used to generate more instructions and responses that follow the requirements stated in the prompt. The machine-generated instruction-following data can be further used to construct with the prompt for in-context-learning in the next data generation iteration. The procedure iterates until a given number of samples are collected. Due to the significantly lower cost and higher response speed of API calls, compared with human annotations, self-instruct is widely used in the research community.

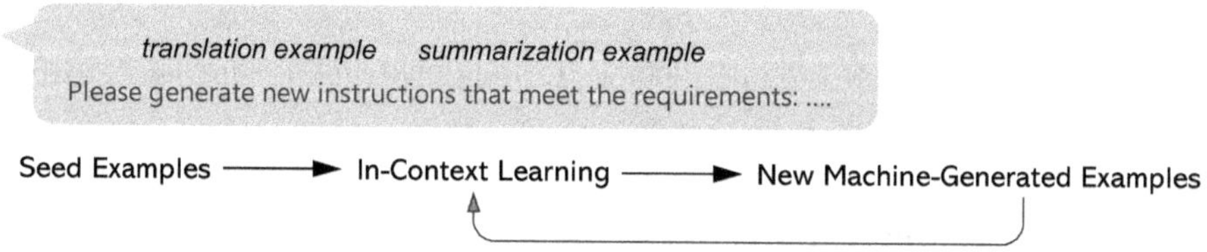

Figure 5.8: Illustration of the self-instruct pipeline [462].

5.2.2 Self-Instruct and Open-Source LLMs

The open-source community has witnessed a surge of open-source LLMs. The success of ChatGPT [333] and GPT-4 [334] offers tremendous opportunities to improve open-source LLMs using instruction-tuning. Figure 5.9 compares several open-source, instruction-tuned LLMs. LLaMA [433] is a series of open-source LLMs, which approximately match the performance of proprietary LLMs such as GPT-3. Self-instruct tuning has been widely adopted to teach LLaMA to follow instructions. To name a few examples, Stanford's Alpaca [425] uses 52K instruction-following samples generated by GPT-3.5; Vicuna [427] uses around 500K high-quality instruction-following samples collected from 150K conversions between human users and GPT [391]; [345] uses GPT-4 as a teacher to generate the responses to Alpaca instructions to significantly lift the SoTA of instruction-tuned LLMs. Many follow-up works [565] improve the instruction-following data for open-source LLMs to achieve a better alignment quality in chat. For a comprehensive review, we refer the readers to [461], where a LLM Tulu is trained on a mix of several high-quality instruction datasets, and a comprehensive comparison is conducted across multiple benchmarks.

	LLaMA	Alpaca	Vicuna	GPT4-Alpaca	...	Tulu
Data Source		GPT-3.5	ShareGPT (Human & GPT)	GPT-4 (text-only)	...	Mixed Data
Instruction-following Data (#Turns)	None	52K	500K (~150K conversions)	52K	...	

Figure 5.9: Model examples of the LLaMA family.

Quick assessment of LLM chatbots. To evaluate the quality of LLM Chatbots, we use *Vicuna-Instructions-80*[1] [427], a dataset with 80 questions that state-of-the-art LLMs [433] find challenging to answer. Besides generic instructions, the instructions fall into 8 categories, including knowledge, math, Fermi, counterfactual, roleplay, generic, coding, writing and common-sense. To quantitatively compare the performance of LLMs, GPT-4 is used to rate each model response using

[1] https://github.com/lm-sys/FastChat/blob/main/fastchat/llm_judge/data/vicuna_bench/question.jsonl

a score from 1 to 10. Despite the simplicity, this evaluation metric turns out to be quite consistent across different settings. The evaluation shows that the open-source LLaMA family performs comparably to SoTA proprietary chatbots.

Further discussions. There are several important topics on LLMs that we have not covered in this section, but are worth exploring in future work.

- *Data-centric AI.* We emphasize that the development of these open-source LLMs is data-centric [304], rather than model-centric. As the training objectives and network architectures are converging, (*e.g.* GPT-like models), the key differential factor is data. For example, the behaviors of the aforementioned LLMs are mainly determined by the instruction-following data.
- *False promise?* It is debatable whether that open-source LLMs could match the proprietary LLMs is a false promise [148]. We argue that the debate is only meaningful if we differentiate two distinctive abilities of LLMs: the ability to understand and follow instructions, and the ability to complete tasks. The former requires an LLM to learn to align its behaviors with human intents while the latter requires an LLM to encode effectively massive knowledge needed to handle various tasks. Instruction-tuned open-source models are good at the former by mimicking ChatGPT's style but perform poorly in terms of factuality in their responses. In [148], the authors conclude that there exists a substantial capability gap between instruction-tuned open-source LLMs and proprietary LLMs which can only be narrowed using an unwieldy amount of instruction-following data or by training a much bigger and more powerful base open-source LLM. Unfortunately, the resource required to train big LLMs is only available in a few industry labs. Thus, with limited compute resource, most academic research labs might focus on two research directions: (1) developing new finetuning methods to better align LLMs with human intent, and (2) exploring new techniques to significantly reduce model training cost.

- *Base LLMs*. Developing powerful LLMs is of great value for both advancing AI research and building commercial AI systems. Based on LLaMA, the open-source community has developed more powerful variants, *e.g.* LLaMA-2, OpenLLaMA [135], MPT [318] and Falcon [343], and released their training recipes [86].

5.3 Instruction-Tuned Large Multimodal Models

In this section, we show how to build an open-source model to mimic multimodal GPT-4, using LLaVA [264] as an example. Another concurrent study is MiniGPT-4 [593].

The research progress in multimodal modeling has often been driven by the advances in NLP in recent years. One successful recipe is to adapt modeling methods originally developed for NLP to vision-and-language tasks. Take self-instruct as an example. The challenge of adapting self-instruct to multimodal modeling is the lack of strong multimodal teachers that are publicly available. Therefore, language models such as GPT-4 have been exploited to create multimodal instruction-following data.

Data Creation

Instead of directly feeding images to OpenAI GPT-4, we use their symbolic sequence representations, as shown in Figure 5.10 (a). In LLaVA, both captions and bounding boxes are used, due to the following reasons: (*i*) They contain complementary information about an image and the combined better represents the image than they do individually; (*i*) Empirical studies show that GPT-4, unlike ChatGPT, can understand captions and bounding box well.

As shown in Figure 5.10 (b), three types of instruction-following data are created: (*i*) multi-turn conversations between users and the model; (*ii*) detailed descriptions that are useful for the model to generate long-form responses; and (*iii*) complex reasoning, to which the responses are more about the implication of the image than content of the image. For example, to answer *"what challenge do these people face?"*, the model first needs to recognize that the image is about a SUV in the parking area and there are luggage on the ground, and infers that the

Context type 1: Captions
A group of people standing outside of a black vehicle with various luggage.
Luggage surrounds a vehicle in an underground parking area
People try to fit all of their luggage in an SUV.
The sport utility vehicle is parked in the public garage, being packed for a trip
Some people with luggage near a van that is transporting it.
Context type 2: Boxes
person: [0.681, 0.242, 0.774, 0.694], person: [0.63, 0.222, 0.686, 0.516], person: [0.444, 0.233, 0.487, 0.34], backpack: [0.384, 0.696, 0.485, 0.914], backpack: [0.755, 0.413, 0.846, 0.692], suitcase: [0.758, 0.413, 0.845, 0.69], suitcase: [0.1, 0.497, 0.173, 0.579], bicycle: [0.282, 0.363, 0.327, 0.442], car: [0.786, 0.25, 0.848, 0.322], car: [0.783, 0.27, 0.827, 0.335], car: [0.86, 0.254, 0.891, 0.3], car: [0.261, 0.101, 0.787, 0.626]

(a) The sequence representation of the image data.

Response type 1: conversation
Question: What type of vehicle is featured in the image?
Answer: The image features a black sport utility vehicle (SUV).
Question: Where is the vehicle parked?
Answer: The vehicle is parked in an underground parking area, likely in a public garage.
Question: What are the people in the image doing?
Answer: The people are trying to fit all of their luggage into the SUV, likely preparing for a trip.
Response type 2: detailed description
The image is an underground parking area with a black sport utility vehicle (SUV) parked. There are three people in the scene, with one person standing closer to the left side of the vehicle, another person in the middle, and the third person on the right side. They are all working together to pack their luggage into the SUV for a trip.
In the parking area, various luggage items are scattered around the vehicle. There are two backpacks, one located near the left rear wheel and the other closer to the right side of the vehicle. Additionally, there are two suitcases, one on the right side of the car and another further away near the center of the parking area. A bicycle can also be seen on the left side of the vehicle.
Other cars are parked around the main SUV, with one car positioned behind it and slightly to the left, another behind and slightly to the right, and the third car further behind on the right side.
Response type 3: complex reasoning
Question: What challenges do these people face?
Answer: In the image, a group of people is standing outside a black SUV in a parking area, surrounded by various pieces of luggage, including suitcases and backpacks. They are facing the challenge of fitting all their luggage into the black SUV. There are multiple suitcases and backpacks to be packed, which suggests that the group has a significant amount of belongings to accommodate. They might have to strategize and arrange the luggage efficiently to ensure that everything fits properly into the vehicle. Additionally, they need to consider the comfort of the passengers and visibility while driving, so the placement of the luggage must not obstruct the driver's view or make the passengers uncomfortable during the trip.

(b) Three types of instruction-following data for the given image.

Figure 5.10: Examples of multimodal instruction-following data. Image credit: [264].

challenge is how to pack luggage into the small truck of the SUV. In total, 158K instruction-following sessions are collected. Note that the self-instruct pipeline for visual instruction-following data collection, as described above, is general. The proprietary GPT-4 model can also be replaced with an open-source model, such as LLaMA-2-70, or even human annotators if budget allows. Whatever tasks one wants the model to perform in the serving stage, it is important to create the corresponding instruction-following data for training.

Network Architecture and Training

As illustrated in Figure 5.11, the network architecture of LLaVA is an instantiation of the general image-to-text generative model framework of Figure 5.2. Specifically, LLaVa is composed of the pre-trained CLIP ViT-L/14 visual encoder [360], the Vicuna LLM [427], and a simple projection matrix (*i.e.*, the linear projection layer). A two-stage instruction-tuning procedure is adopted to train the model. (*i*) *Stage 1 is pre-training for feature alignment*, where only the projection matrix is updated based on a subset of CC3M [50]. (*ii*) *Stage 2 is end-to-end finetuning*, where both the projection matrix and LLM are updated based on the curated multimodal instruction-following data collected from real-world, user-oriented applications.

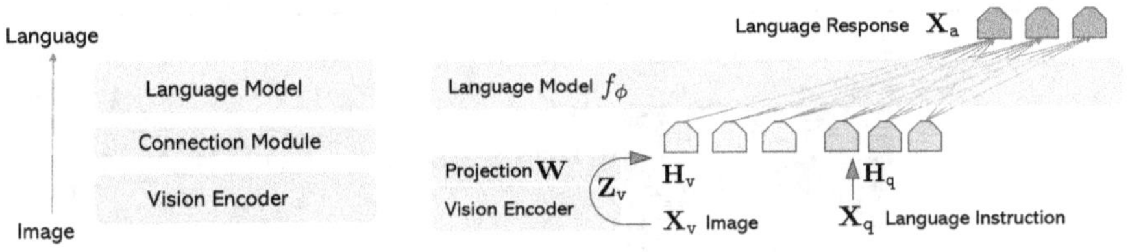

Figure 5.11: Network architecture: Left: General LMM; Right: LLaVA. Image credit: [264].

Performance

Visual chat: towards building multimodal GPT-4 level chatbot. LLaVA is finetuned on the model-generated multimodal instruction-following data, which contains a diverse set of task instructions and responses for daily, user-oriented applications. It is empirically observed that finetuning the linear projection layer only is sufficient for the chat demo/scenarios, although it requires a larger number of training iterations. To evaluate model performance, the LLaVA-Bench benchmark is constructed. It consists of two datasets: (*i*) LLaVA-Bench (COCO) contains 30 unseen COCO images with 90 new language-image instructions; and (*ii*) LLaVA-Bench (In-the-Wild) contains 24 images with

60 questions. Each image can be associated with three types of instructions: conversation, detailed description and complex reasoning. The ground-truth answers are collected by manually re-writing GPT-4 outputs. We test LLaVA and use language-only GPT-4 to rate their responses from score 1 to 10. Overall, LLaVA achieves 85.1% relative score compared to ground-truth on LLaVA-Bench (COCO), and 73.5% on LLaVA-Bench (In-the-Wild). On the latter dataset, Google Bard (July 19, 2023) and Microsoft BingChat (June 29, 2023) achieves 77.8% and 71.5%, respectively. It indicates the effectiveness of the proposed self-instruct method for multimodal modeling. One example is shown in Table 5.1.

Table 5.1: Example prompt comparing LLaVA, GPT-4, BingChat, Bard, MiniGPT-4, BLIP-2, and OpenFlamingo's visual reasoning capabilities in understanding the humor. LLaVA and GPT-4 both explain the meme and its humor, while GPT-4 produces a more concise answer. Table credit: [264].

Visual input example, Chicken Nugget Map:	

User	Can you explain this meme in detail?
LLaVA	The meme in the image is a creative and humorous take on food, with a focus on chicken nuggets, as the center of the universe. The meme begins with the phrase "Sometimes I just look at pictures of the Earth from space and I marvel at how beautiful it all is..." The main image features a tray of cooked chicken nuggets, with the chicken pieces positioned to resemble the continents and islands on a world map. The meme playfully suggests that the chicken nuggets represent the Earth, and the various locations depicted in the photo are actually chicken nugget versions of different places. The punchline of the meme appears in smaller text at the bottom, reading "I mean, it's not the real Earth, but how beautiful it is all is." This lighthearted and entertaining meme combines food and humor, using the image of chicken nuggets to create a funny and imaginative representation of the world.

Table 5.1: Continued.

User	Can you explain this meme?
GPT-4	This meme is a joke that combines two unrelated things: pictures of the earth from space and chicken nuggets. The text of the meme suggests that the image below is a beautiful picture of the earth from space. However, the image is actually of chicken nuggets arranged to vaguely resemble a map of the world. The humor in this meme comes from the unexpected juxtaposition of the text and the image. The text sets up an expectation of a majestic image of the earth, but the image is actually something mundane and silly.
User	Can you explain this meme in detail?
Bing Chat	Sure, I can try to explain this meme in detail. This is a **chicken nugget meme**, which is a popular online meme that typically features a picture of a chicken nugget with a caption that is funny or relatable. The meme has been popular since the early 2000s, and continues to be one of the most popular memes today. This particular meme is based on a contrast between the text and the image. The text above the image reads "Sometimes I just look at pictures of the earth from space and I marvel at how beautiful it all is." This is a sentiment that many people can relate to, as looking at the earth from space can inspire awe and wonder. However, the image below the text shows a baking tray with chicken nuggets arranged in the shape of the world map. This is a humorous twist, as chicken nuggets are not usually associated with beauty or marvel, but rather with fast food or junk food. The image also implies that the person who made this meme loves chicken nuggets so much that they see them everywhere, even when looking at pictures of the earth from space. The meme is meant to be funny and absurd, and to poke fun at the person's obsession with chicken nuggets.I hope this explanation helps you understand this meme better.

Table 5.1: Continued.

User	Can you explain this meme in detail?
Bard	Sure. The meme you sent me is a humorous take on the idea of looking at pictures of the earth from space and marveling at its beauty. The text in the meme says "Sometimes I just look at pictures of the earth from space and I marvel at how beautiful it all is." However, the image shows a map of the world made out of chicken nuggets. This juxtaposition of the text and the image is what creates the humor in the meme. The text suggests that the speaker is looking at something beautiful and awe-inspiring, but the image shows something that is more mundane and humorous. The meme also plays on the fact that chicken nuggets are often seen as a junk food. By depicting a map of the world made out of chicken nuggets, the meme suggests that even something as mundane as chicken nuggets can be beautiful, or at least, more beautiful than we might think. Overall, the meme is a humorous way to poke fun at the idea of finding beauty in the mundane. It also shows how our perception of beauty can be influenced by our expectations. Here are some other interpretations of the meme: (1) The meme could be seen as a commentary on the way we consume images of the earth from space. We often see these images as representations of beauty and perfection, but the meme reminds us that the earth is also a messy and imperfect place. (2) The meme could also be seen as a critique of the fast food industry. The use of chicken nuggets to represent the world suggests that we are increasingly seeing the world through the lens of consumerism. Ultimately, the meaning of the meme is up to the interpretation of the viewer. However, it is clear that the meme is a humorous and thought-provoking way to explore the idea of beauty.
User	Can you explain this meme in detail?
MiniGPT-4	This meme is a picture of a pan full of fried chicken. The chicken is cooked and appears to be ready to be served. The caption reads, "Sometimes it's just like looking at pictures from space and marveling at how beautiful it all is." This meme is likely meant to be humorous and playful, as it compares the beauty of space to the simple pleasure of eating fried chicken. It is likely meant to be a lighthearted and funny way to express appreciation for both the wonders of space and the simple pleasures of food.
User	Can you explain this meme in detail?
BLIP-2	sometimes i just look at pictures of the earth from space and marvel how beautiful it is
User	Can you explain this meme in detail?
OpenFlamingo	It's a picture of a chicken nugget on the International Space Station.

Science QA: New SoTA with the synergy of LLaVA with GPT-4. LLaVA is finetuned on a multimodal reasoning dataset in the science domain [290]. LLaVA alone achieves 90.92% in accuracy. We further explore using language-only GPT-4 as the judge to predict the final answer based on its own previous answers and the LLaVA answers. This "GPT-4 as judge" scheme yields a new SoTA of 92.53%.

OCR in the wild: An emergent property. LLaVA has never been explicitly trained on OCR data, *i.e.* images that contains scene text that describes the corresponding caption. Surprisingly, the model shows strong ability of zero-shot OCR task transfer in the wild.

5.4 Advanced Topics

The recent progress of instruction-tuned LMMs is illustrated in Figure 5.12 (a). Due to the popularity of ChatGPT and GPT-4, instruction-tuned LMMs appear as an emerging line of research in the past three months since the release of GPT-4. Alpaca [425] and Vicuna [427] are instruction-tuned LLMs based on LLaMA. In two weeks, MiniGPT-4 [593] and LLaVA [264] extend Vicuna to multimodal and can chat with human users in a vision-language setting. In ten days, LLaMA-Adapter v2 [131] and mPlug-OWL [522] are proposed to compare performance with MiniGPT-4 and LLaVA, indicating the beginning of model evolution. The data points in April are relatively sparse. In May and June, a large number of LMM papers are published on arXiv, which significantly advance the research.

It is difficult to track all the recent papers in our literature review. To better organize the literature, In what follows, we group some recent papers based on specific research topics, as shown in Figure 5.12 (b). These papers present both (proprietary) LMMs with billions of parameters include GPT-4 [334], Flamingo [4], PaLM-E [108] and KOSMOS-1 [183], and light-weighted, open-source models such as LLaVA and MiniGPT-4. The other important topics that we are not going to discuss here include extensions of RLHF [150], dense prediction [55], [456], [541], video [237], [297], [555], image generation [207] and embodied agent [322].

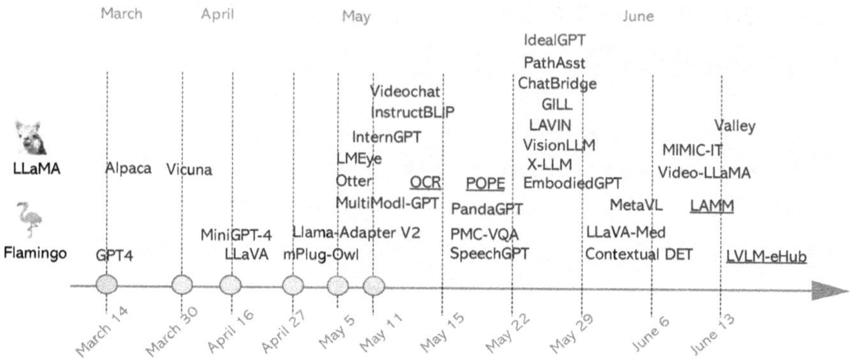

(a) The surge of papers on LMMs from March 14, 2023 to June 19, 2023. To date, an increasing number of related papers in this line of research have been appearing. Those with an underline indicate benchmarks, otherwise indicate models.

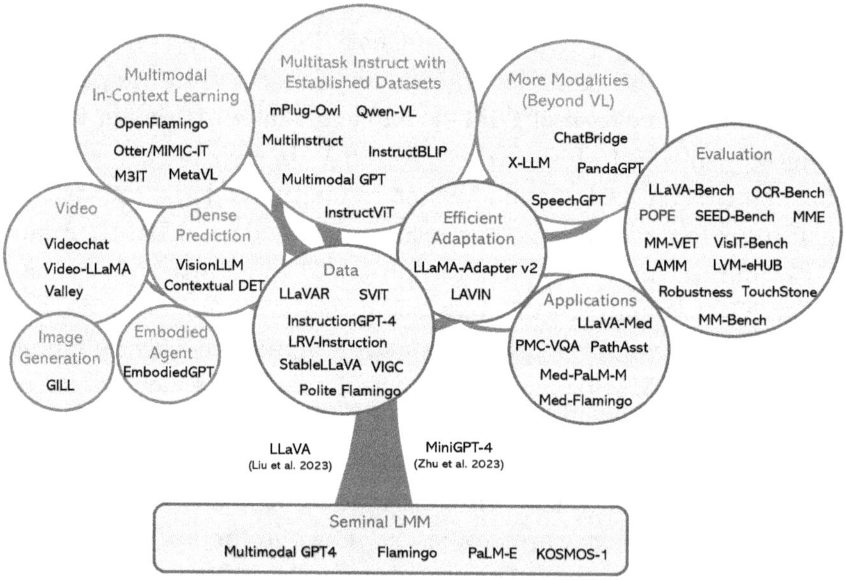

(b) Summary and categorization of papers on LMMs.

Figure 5.12: Review and summary for the emerged LMM literature. Due to space constraints, some methods are not displayed visually, but we aim to address them in the accompanying text.

More Modalities (Beyond Vision and Language)

While LMM extends LLM by adding the vision modality, several attempts have been made to extend the framework to include other modalities beyond vision and language. PandaGPT leverages ImageBind to add more modalities into LMMs. The ImageBind model [139] learns a single, shared representation space for text, image, video, audio and sensors that record depth (3D), thermal (infrared radiation), or inertial measurement units (IMU), which calculate motion and position. ImageBind provides a holistic understanding of the visual world that connects objects in a photo with how their sound, 3D shape, temperature, and movement. By training a projection layer for one modality in LMM, the model can zero-shot transfer to other modalities, thanks to the shared multimodal embedding space. Another representative model is SpeechGPT, where the language and speech modalities are allowed in both inputs and outputs. Despite rich model variations, the idea to connect diverse modalities is similar to connecting images to texts in vision-language models. NExT-GPT [481] connects an LLM with multimodal adaptors and different diffusion decoders, allowing NExT-GPT to perceive inputs and generate outputs in arbitrary combinations of text, images, videos, and audio. The LMM framework has also been successfully extended to speech [578], 3D [175], [464], and point cloud [498].

Improving Visual Instruction Data Quantity and Quality

Given that model architectures are converging to that of GPT, the performance of LMMs is primarily determined by training data. Therefore, it is crucial to improve the quantity and quality of visual instruction tuning data. SVIT [576] extends the LLaVA data generation pipeline to include region description to prompt GPT-4, in addition to the caption and box data as shown in Figure 5.10 (a). The dataset is scaled up to 3.2 million, which is 20 times larger than the dataset used in LLaVA.

Unlike existing studies that primarily focus on positive instruction samples, LRV-Instruction [261] includes both positive and negative instructions for robust instruction-tuning. Other examples along this line include LLaVAR [573] that adds OCR-related instruction-tuning data for text-rich image understanding, and StableLLaVA [247] that uses model-synthesized images as image-dialogue data. Polite Flamingo [53] trains LLM to re-write instruct data. Instead of leveraging GPT-4 for data generation, VIGC [442] uses LMMs to generate instruction-tuning data and progressively enhances its quality on-the-fly. Similar to LIMA [585] for NLP, InstructionGPT-4 [472] shows that the quality of the instruction-tuning data is more important than its quantity. A better version of MiniGPT-4 is finetund using 200 high-quality samples (6%), selected from the 3500 samples used to finetuned the original MiniGPT-4.

Multitask Instruct with Established Academic Datasets/Tasks

As discussed in Section 5.2, instruction tuning in the language domains is implemented in two different ways: finetuning a language model on a wide range of tasks using human-annotated prompts and feedback [338], or supervised finetuning using public benchmarks and datasets augmented with manually or automatically generated instructions [463]. The former works well for user-oriented daily life tasks, and the latter achieves good performance on established benchmarks. LLaVA and MiniGPT-4 fall into the former. Other works use either the latter, such as MultiInstruct [500], or combine both, such as mPlug-OWL [522], InstructBLIP [90], Multimodal-GPT [142], Instruction-ViT [486] and Qwen-VL [20].

For example, InstructBLIP combines chat and benchmark instruction-following data. As shown in Figure 5.13, InstructBLIP transforms 26 publicly available datasets, covering a wide variety of tasks and capabilities, into the same instruction tuning format. Trained on 13 held-in datasets, InstructBLIP obtains SoTA zero-shot performance across all 13 held-out datasets, substantially outperforming the BLIP-2 and Flamingo models. Qwen-VL scales up both image-text pair data for pre-traning and academic datasets for multi-task pre-traning, and achieves good performance on many tasks.

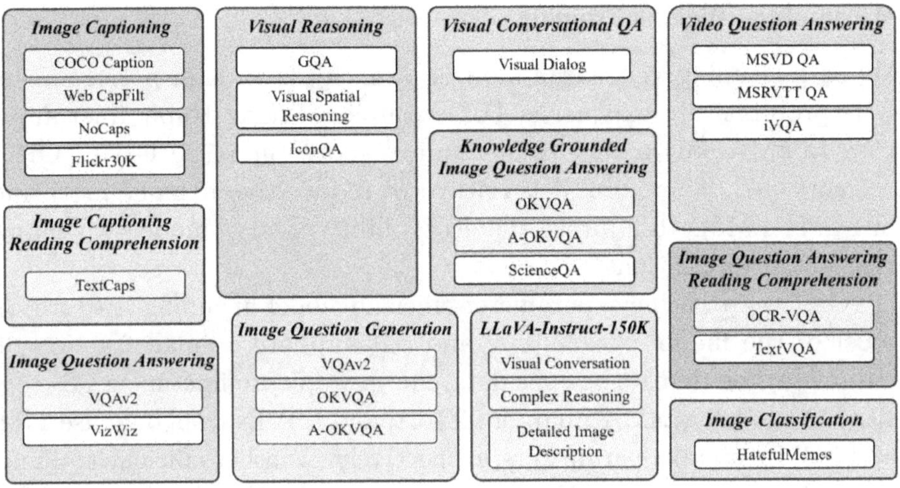

Figure 5.13: The vision-language tasks covered in InstructBLIP. Image credit: [90].

Multimodal In-Context-Learning

In-context learning of LLMs refers to the LMM's capability of performing new tasks by prompting the model with task-specific examples and a task description. Similarly, multimodal in-context-learning refers to a visual and text interface (prompting) that can instruct an LMM to solve a multimodal task. Given a few examples of visual-input-text-response pairs in a multimodal prompt, the LMM can be instructed to answer questions regarding input images or videos. The studies that explore how to extend in-context-learning from language to multi-modalities include OpenFlamingo [15], Otter [224], M^{3}IT [238], MetaVL [313] and Sparkles [186].

OpenFlamingo [15] is an open-source version of DeepMind's Flamingo model, trained on Multimodal C4 dataset [596], which consists of a billion interleaved image-text samples. To explicitly enhance the multimodal in-context-learning ability of LMMs, the MIMIC-IT dataset [223] is constructed. It contains 2.4M multimodal instruction instances with in-context examples. Tuning OpenFlamingo on MIMIC-IT leads to the Otter model, which shows strong instruction-following ability.

Parameter-Efficient Training

While finetuning very large models often leads to high performance, it is prohibitively expensive; For example, regular 16-bit finetuning of a LLaMA-65B model [433] requires more than 780 GB of GPU memory [97]. Therefore, it is critical to reduce the memory footprint of LLMs/LMMs to improve the accessibility of large models for many real-world applications.

One way to achieve parameter-efficient model fineuning is to freeze most of the model parameters, and only update a small fraction of parameters on domain-specific data. For example, to finetune a 7B/13B LLaMA model, LLaMA Adapter v2 [131] and LAVIN [295] only use 14M and 3.8M trainable parameters, respectively. Another efficient training method is quantization. QLoRA [97] finetunes a 65B LLaMA model for 24 hours on a single GPU, achieving 99.3% of the performance level of ChatGPT. Since instruction tuning typically uses a small amount of data, parameter-efficient training and model quantization are widely used. Both LoRA [176] and QLoRA are supported in the LLaVA codebase. It is empirically shown in [293] that LoRA/QLoRA can achieve similar performance with full-model tuning when scaling LLaVA to 33B and 65B. The study uses a 150K instruct dataset and evaluates models on LLaVA-Bench.

Benchmarks

While LMMs have shown excellent visual recognition and reasoning capabilities in an open-set manner with free-form text across many scenarios, the evaluation of LMMs remains challenging. Several benchmarks have been developed to evaluate various aspects of LMMs, ranging from specific abilities including OCR [279], hallucination (POPE [250] and HaELM [449]) and adversarial robustness [579], to comprehensive evaluation such as LAMM [526], LVLM-eHub [497]. We summarize the LMM evaluation benchmarks in Table 5.2. Among them, LLaVA-Bench is the first open-world visual chat benchmark, designed specifically for evaluating LMMs. Recently, multimodal experiments have been conducted to compare open-source LMMs with commercial models such as BingChat and Bard on LLaVA-Bench [264] and LVLM-eHub [390].

Table 5.2: Comparisons of recently proposed LMM evaluation benchmarks.

Benchmark	Capability to Evaluate	Statistics	Metric
LLaVA-Bench [264]	Multi-turn QA, detailed description, reasoning	Two subsets: 90 samples on COCO and 60 samples on In-the-Wid	Relative score via GPT-4 evaluation
OCR-Bench [279]	Zero-shot OCR	A suite of 23 OCR-related academic tasks	Accuracy
MMBench [278]	Perception (coarse, fine-grained single-instance and cross-instance) and Reasoning (attribute, relation, logic)	2974 multiple-choice samples in 20 ability dimensions	Circular evaluation via Chat-GPT answer extraction
M3Exam [569]	Multilingual, multimodal, and multi-level assessment	12,317 questions in 9 languages, with 2,816 questions involving images	Accuracy on multiple-choice questions
MME [123]	Perception and Cognition	14 tasks	Accuracy on "yes" or "no"
LAMM [526]	Various 2D/3D vision tasks	9 image task with 62K samples, and 3 point cloud tasks with 12K samples	Traditional CV task metrics
LVLM-eHub [497]	six multimodal capabilities such as VQA and embodied AI	47 standard text-related visual benchmarks	CIDEr and accuracy; Arena with human judgment
SEED-Bench [225]	Comprehension of both the image and video modality	19K multiple choice questions in 12 dimensions	Accuracy on multiple-choice questions
VisIT-Bench [30]	Real-life vision-language instructions	592 samples in 70 tasks	Elo, matches
MM-VET [537]	Integrated capabilities in recognition, OCR, spatial, knowledge, math, language	200 samples	GPT-4 evaluation
TouchStone [21]	Five abilities: basic description, visual recognition, visual comprehension, visual storytelling, and multi-image analysis	908 dialogues in 27 tasks	GPT-4 evaluation
SciGraphQA [244]	Scientific graph question-answering	3K test samples	CIDEr, BLEU-4, and ROUGE

It is surprising that LMMs show strong zero-shot OCR performance in the wild, without explicitly training on text recognition data. To shed light on how this might happen, a comprehensive empirical study [279] is conducted to compare open-source LMMs on 24 academic text recognition datasets, as shown in Figure 5.14. Three observations are reported: (*i*) LLaVA consistently outperforms MiniGPT-4 on 21 out of 24 datasets, despite that the training data in LLaVA is an order of magnitude smaller. (*ii*) Training with significantly more training data improves OCR performance, as demonstrated by BLIP2 [234] and mPLUG-Owl. (*iii*) In most cases, supervised learned models significantly outperform zero-shot LMMs. However, it is worth noting that in the WordArt dataset [488], which contains challenging artistic texts, BLIP2 surpasses the supervised SoTA results. This reveals the potential of LMMs in recognizing complex text types.

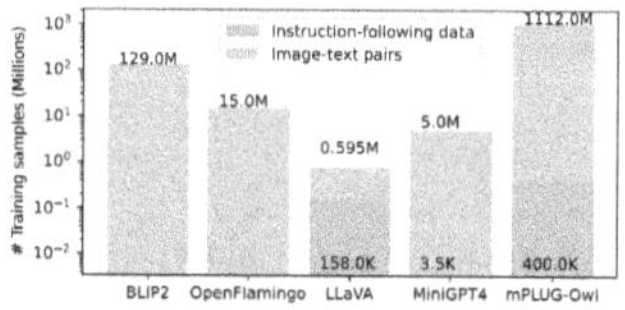

	14 tasks: OCR	10 tasks: OCR + Reasoning
BLIP-2 $OPT_{6.7b}$	62.12	7.31
BLIP-2 $FlanT5_{XXL}$	63.74	13.59
OpenFlamingo	56.68	11.28
LLaVA	53.30	10.37
MiniGPT4	40.12	7.82
mPLUG-Owl	61.79	16.03
Supervised-SoTA	**84.83**	**72.75**

Figure 5.14: Zero-shot OCR performance of LMMs on 24 datasets. Image credit: [279].

Applications

The success of ChatGPT/GPT-4 in general domains has spurred the development of AI assistants in the vertical domains such as medicine, gaming, education. Such domain-specific assistants have several advantages over their general domain counterpart: (*i*) training with high-quality domain-specific data improves model's performance on specific tasks; (*ii*) the model size can be smaller, and thus these models are more cost-efficient for serving; and (*iii*) sensitive user prompt data can be stored locally to protect privacy.

To improve the text recognition ability of LMMs, OCR-specific models have been developed, including BLIVA [180], LLaVAR [573], mPlug-DocWL [519]. LMMs have also been recently explored in the biomedical domain [228], [418], [570], where conversational AI agents have already shown the potential of assisting biomedical practitioners in real-world scenarios. LLaVA-Med [228] is a cost-efficient, vision-language, conversational assistant that can answer open-ended research questions about biomedical images. LLaVA-Med is developed by (1) collecting a large-scale, broad-coverage biomedical figure-caption dataset extracted from PubMed Central, (2) using GPT-4 to self-instruct open-ended instruction-following data from the captions, and then (3) finetuning LLaVA using a curriculum learning method. Specifically, the model first learns to align biomedical vocabulary using the image-caption pairs, then learns open-ended conversational semantics using GPT-4 generated instruction-following data. The two-stage learning process mimics how a layperson gradually acquires biomedical knowledge. Figure 5.15 presents a few examples of the biomed visual conversations with different chatbots. LLaVA-Med behaves like an expert and is able to answer the questions that require biomedical knowledge. In contrast, LLaVA hallucinates a lot, like a layman. LLaVA-Med is instrumental to the development of several generalist biomedical AI models, including Google Med-PaLM-M [435], Stanford Med-Flamingo [315] and radiology generalist [476].

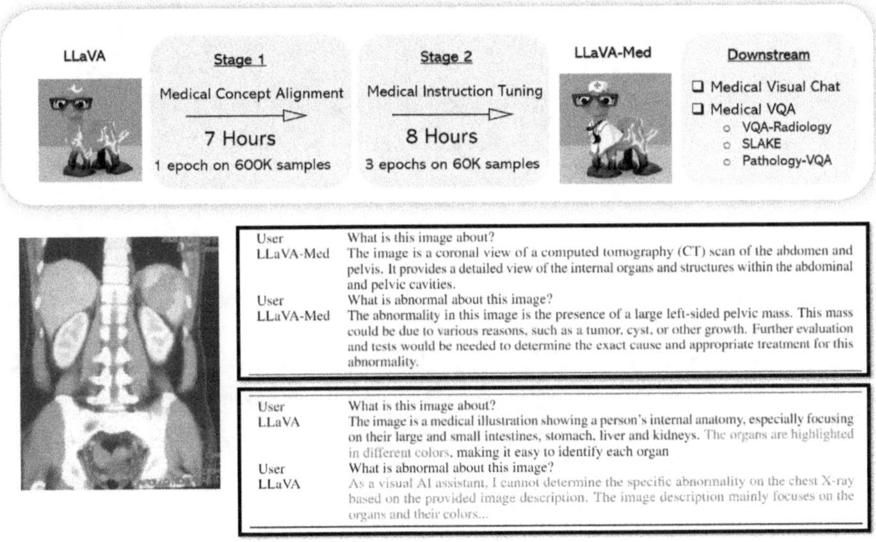

Figure 5.15: Application of LMMs to biomedical images. Top: Domain adaptation from LLaVA to LLaVA-Med. Bottom: The chat behaviors of two chatbots. Image credit: [228].

5.5 How Close Are We To OpenAI Multimodal GPT-4?

With all the works described above, are we close to (or, even surpassing) the OpenAI Multimodal GPT-4? It is encouraging to see that the open-source community has quickly developed a variety of models and prototypes with various new capabilities, laying a foundation for developing general-purpose multimodal AI agents. For example, LLaVA/Mini-GPT4 can reproduce the results generated by OpenAI GPT-4. CM3leon [533], Emu [416] and GILL [207] extend LMMs for end-to-end image generation, which can not be achieved using multimodal GPT-4.

However, there is still a visible gap between open-source models and GPT-4 in model performance. It is not difficult to find examples where GPT-4 can give correct answers but open-source models, such as LLaVA, cannot. One possible reason is that GPT-4 is much bigger in model size and is trained on much larger corpora. Thus, one important area for future research is cost-efficient modeling.

6

Multimodal Agents: Chaining Tools with LLM

Large Language Models (LLMs) [85], [334] have shown emergent abilities to generalize to new domains and scenarios via in-context learning. Thus, there is a growing interest in exploring a new modeling paradigm of building general-purpose AI agents that shifts from building standalone models for solving a finite set of pre-defined problems to synergistically chaining multiple tools or experts with LLMs to solve sophisticated, open-world problems. Unlike what has been discussed in Section 5, such an AI system can be built without any large-scale model training involved, but based on prompting engineering where a task description and a few demonstration examples are offered in a prompt to instruct the LLM to compose existing tools to generate answers. Figure 6.1 shows the AI robot that works with improved efficiency by streamlining tool use in a manufacturing environment.

In this section, we review the fast-evolving literature on building *multimodal agents* that can chain different multimodal experts or tools with LLMs to solve complex multimodal understanding tasks. We start with an overview on the evolution of this modeling paradigm in Section 6.1, highlighting the difference between traditional approaches and the new modeling paradigm of chaining tools with LLMs. Section 6.2

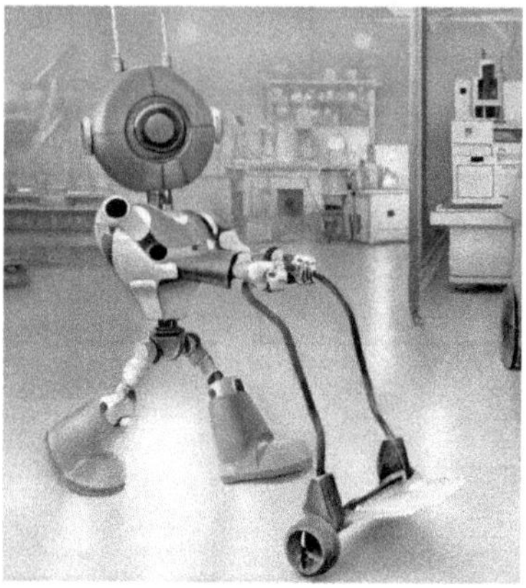

Figure 6.1: The AI-powered assistant that brings a new level of efficiency to work with tools in a manufacturing environment full of equipments.

gives an overview of multimodal agents. Using the MM-ReAct [513] agent as an example, Section 6.3 describes in detail how to build a multimodal agent, its emergent capabilities in multimodal understanding, and how it can be extended to incorporate LLMs and other tools. Finally, Section 6.4 discusses advanced topics on how to improve and evaluate multimodal agents, and the diverse applications powered by multimodal agents.

6.1 Overview

We first present the evolution of modeling paradigms, from task-specific models to large multimodal models. We then describe the new modeling paradigm of chaining tools with LLMs, where instead of training models from scratch, pre-trained LLMs can be augmented by many pre-built tools (via APIs) to construct multimodal agents.

Evolution of modeling paradigm. As summarized in Figure 6.2, we are witnessing the transition from task-specific models towards general-purpose assistants across language, vision, and multimodal research.

We started with **task-specific models** that are trained on small-scale well-annotated data. This results in dedicated models [7], [239], [538] for each task or even each dataset.

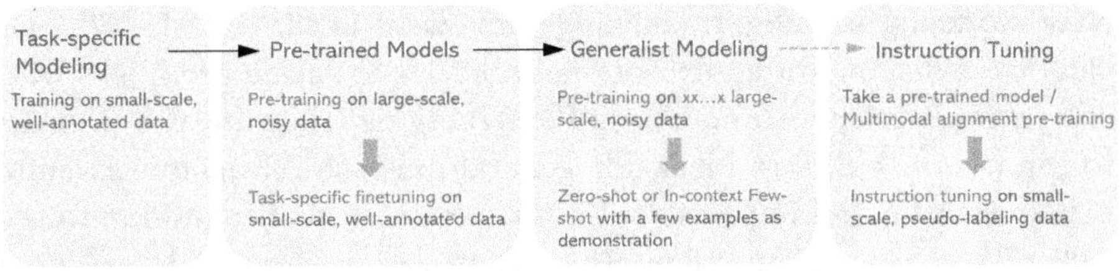

Figure 6.2: Evolution of modeling paradigm.

We then transitioned to the phase of **pre-trained models**, where the pretrain-then-finetune paradigm has been widely adopted across both NLP and vision-language (VL) research. During pre-training, the model can take advantages of large-scale, web-crawled noisy data, for example, millions to billions of image-text pairs [79], [446], or billions of text tokens [98], [277]. However, task-specific finetuning is still required, similar to the case of task-specific models. This paradigm has led to many well-known models, such as BERT [98], RoBERTa [277] and DeBERTa [167] in NLP, and UNITER [79], OSCAR [246] in VL. These early VL foundation models were considered large-scale (trained with 10M image-text pairs), but could be perceived as intermediate or even small scale in today's view (billions of pairs).

Now, we are entering a new era of **generalist modeling**, where model pre-training has been further scaled up to trillions of text tokens [131]. For downstream adaptation, these generalist models have shown strong performance with in-context learning using a few demonstration examples, or even zero-shot evaluation. These models are what we now refer to as large language/multimodal models, including the GPT family [333], [334], PaLM family [85], [108], LLaMa [433], and Flamingo [4].

Based on the generalist models, the pipeline of building **instruction-following models**, as described in Section 5, follows the pretrain-then-finetune paradigm. For example, Alpaca [425] is initialized using the pre-trained LLaMa [433], then finetuned on a small-scale instruction tuning dataset. Similarly, for instruction-following VL models (*e.g.* LLaVA [228]), an additional stage of image-text alignment pre-training is introduced to align the visual representations to the language representations, followed by visual instruction tuning.

New modeling paradigm: chaining tools with LLM. LLMs [37], [85], [334] have demonstrated exceptional abilities to tackle new tasks with only a few examples and textual instructions, showing the promise to serve as a building block for general-purpose AI agents. Despite being versatile and impressive, LLMs encounter challenges due to the lack of some basic functionalities, such as mathematical reasoning and information retrieval. Furthermore, a fundamental limitation of LLMs is that they only represent the world described by their training data, which inevitably becomes outdated over time. Re-training LLMs regularly with the latest information is simply not feasible.

Meanwhile, many tasks with real-world impact cannot be readily tackled by LLMs alone. For example, accessing up-to-date information and performing computations can be done via existing tools (*e.g.* a search engine or calculator). Hence, recent research has explored a new modeling paradigm by augmenting LLMs with external NLP tools [3], [185], [326], [344], including calculators, search engines, translation systems, calendars, or even API calls on other models.

LLM augmentation studies mainly focus on a single modality, *i.e.*, language, in which the output of the tools are in text format, which can be readily fed to LLMs as additional knowledge. However, we live in a multimodal world and a truly intelligent agent should be able to perform multimodal reasoning and actions. How to equip LLMs with perception of multimodal signals by plugging in other tools is the focus of the rest of the section.

6.2 Multimodal Agents

There are several representative works on building multimodal agents using tools or vision experts, including VISPROG [154], Visual ChatGPT [477] and MM-ReAct [513]. VISPROG is one of the first works on using programming language to chain different vision tools with a LLM. Visual ChatGPT enables dialogue-based image editing by augmenting ChatGPT [333] with various image generation tools. MM-ReAct shows that collaborating various advanced vision experts allows ChatGPT to perform complex multimodal reasoning tasks. Figure 6.3 presents the fast-evolving literature on multimodal agents from November 18,

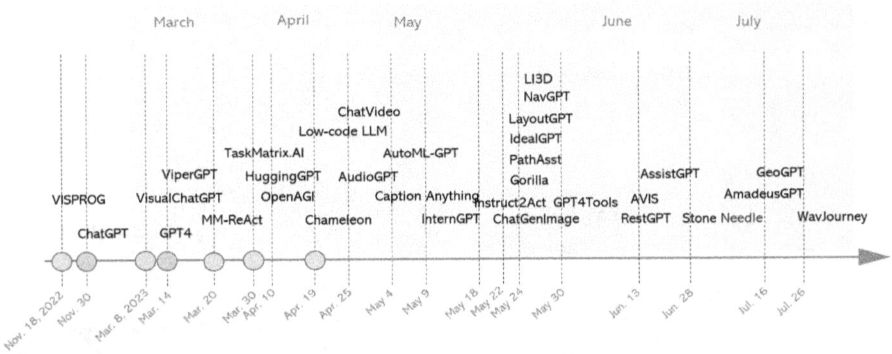

Figure 6.3: The surge of multimodal agents from November 18, 2022 to July 26th, 2023.

Table 6.1: Glossary of representative works on chaining tools with LLMs. The LLMs used in these works include GPT-3 [37], GPT-J [441], ChatGPT [333], GPT-3 Codex [59] and GPT-4 [334]. [search engine icon]: search engine. [code icon]: code. [API icon]: APIs other than search engine. [open-source icon]: open-source models.

Model	LLM	Tools	Tool Size	Tool Execution	Multimodal
ART [340]	GPT-3	search engine, API, code	3	Program	✗
Toolformer [384]	GPT-J	search engine, API	5	Natural language	✗
VISPROG [154]	GPT-3	code, open-source	> 10	Program	✓
Visual ChatGPT [477]	ChatGPT	open-source	> 10	Natural Language	✓
ViperGPT [419]	GPT-3 Codex	API, open-source, code	> 10	Program	✓
MM-ReAct [513]	ChatGPT/GPT-4	search engine, API	> 10	RegExp Match	✓
HuggingGPT [396]	ChatGPT	open-source	> 10	Natural Language	✓
Chameleon [291]	GPT-4	search engine, open-source, code	> 10	Natural Language	✓

2022 to July 26th, 2023. We also present some exemplary multimodal agents in Table 6.1, which includes two representative studies in the NLP domain.

A typical multimodal agent framework is illustrated in Figure 6.4. A user directly interacts with **Tool Allocator**, which functions as the *brain* of the AI agent. In most state-of-the-art multimodal agents, the tool allocator is often implemented using a LLM. To achieve the user's goal, an LLM-powered multimodal agent needs to figure out all the necessary steps (planning), activate corresponding tools for execution, and generate the final response based on the execution results. Next, we describe the three key components of multimodal agents.

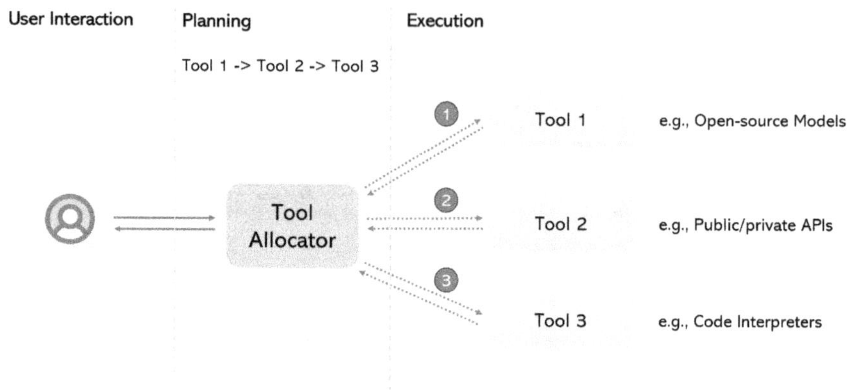

Figure 6.4: An overview of multimodal agent framework.

Tools. Tools are external modules that can be activated by the multimodal agent (or its LLM) to perform (sub-)tasks that cannot be accomplished by the LLM alone. These tools can be open-source models, public or private APIs, or code interpreters. For example, as the LLM only accepts language inputs, a multimodal agent often calls tools (*e.g.* a vision encoder) to process multimodal inputs.

Planning. During planning, the LLM decomposes a user task into smaller, manageable sub-tasks, and makes a step-by-step execution plan, where each step involves activating one or more external tools. There are two ways to teach LLMs for planning. One is to prompt the LLM with in-context few-shot examples of using candidate tools. This approach can easily extend to a growing set of Tools. But there is a limit on how many tools can be used in one step due to the input length limit of LLMs. The other approach uses large amounts of annotated data to finetune the LLM. There is no length limit, but the finetuned LLM might not be as robust and generalizable as it was before the finetuning.

Execution. To execute the generated plan, the agent (or its LLM) activates the tools using three methods: (1) regular expression matching [513], (2) directly prompting the LLM to generate executable programs [419], or (3) leveraging in-context few-shot learning capability of the LLM by providing few-shot examples and a natural language instructions that describe the role of each tool [291]. The execution results of the tools are aggregated by the LLM to generate the final response to the user.

6.3 Case Study: MM-ReAct

We use MM-REACT [513] as a case study to show how to build a multimodal agent, its emerging capabilities in multimodal understanding, and how it can be easily extended to incorporate the latest and strongest LLMs and tools.

6.3.1 System Design

MM-ReAct [513] is composed of ChatGPT and a number of multimodal tools.[1] ChatGPT is the planner for multimodal reasoning, and it is augmented with multimodal tools to support both inputs and outputs in multi-modalities, including text, image, and video, as shown in Figure 6.5.

Figure 6.5: Input/Output modalities of MM-REACT [513].

[1]These tools are referred to as experts in [513].

Figure 6.6 shows the system design of MM-ReAct. The **multimodal tools** in MM-ReAct are mainly computer vision models that take an image as input and represent the image content from different perspectives. For instance, an image captioning model generates a natural description of the image; an OCR model extracts the scene text in the image; a celebrity recognition model identifies the celebrity names; and an object detection model extracts the salient objects with bounding box locations. ChatGPT serves as the brain of the agent. Given a user query, ChatGPT generates a plan to determine which tools to use and in what order. Next, we explain planning and execution of MM-ReAct using the example in Figure 6.6.

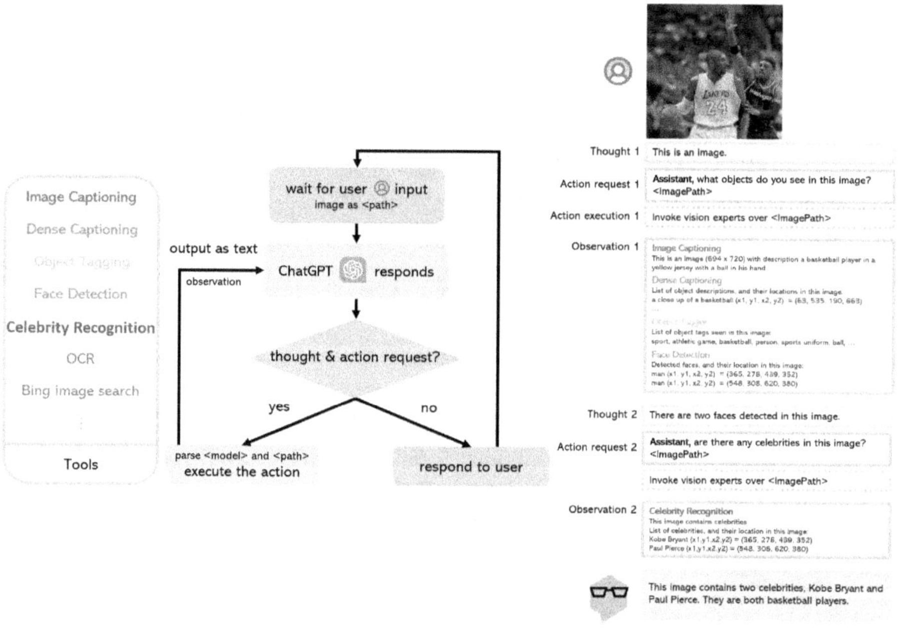

Figure 6.6: System design of MM-ReAct [513].

User prompt. As ChatGPT only accepts language inputs, to process image inputs, we simply use the path of an image file as the input to ChatGPT. The file path is used as a placeholder, allowing ChatGPT to activate other tools (*e.g.* image understanding models) to process the image after planning. Besides images, a user can also input intent in text format (*e.g.* a question about the input image). When there is no text input from the user, the user intent is assumed to obtain an overall description of the image.

Planning. Upon receiving the input image and user prompt, ChatGPT determines what tools to use. Inspired by REACT [517], MM-REACT instructs ChatGPT to respond with certain watchwords, such as *"Assistant, what objects are there in the image? <file path>"*, if a specific tool is required (i.e., *action request* in Figure 6.6). In practice, users can tell whether a multimodal tool is needed by simply string-matching the keyword "Assistant," in ChatGPT's response.

In MM-ReAct, ChatGPT is configured to reveal the *thought* (reasoning) process to show why an external tool is used, which has been proven beneficial for NLP tasks [517]. In addition, to properly activate a tool, both instructions and in-context examples are added as the prefix in the prompt to ChatGPT. Each tool is described by its model name, a general description of its capability, the input data format, and the output information. After describing each tool, a few in-context examples are also included in the prompt to improve the model output.

Execution. Given the action request from ChatGPT, the tool name and the file path are parsed via regular expression matching to invoke the tool (*action execution*).

Consider the example shown in Figure 6.6. Upon receiving the input image from a user, ChatGPT first invokes a series of tools to understand the image. The invoked tools include image captioning to generate an overall description of the image; dense captioning to get a region-level, detailed description about the objects in the image; object tagging to get the tags of the objects in the image; and face detection to get the box coordinates of the two faces mentioned in the object tags. The outputs from these tools (observations) are converted into a text string, and fed back to ChatGPT.

Combining observations with chat history, ChatGPT can invoke additional tools or return the final answer to the user. In this specific example, ChatGPT performs a second round of *thought-action-observation* over the two faces detected in the image and calls celebrity recognition to obtain the names of these two persons.

Response generation. When ChatGPT decides that no external tools are needed, it gathers all observations and summarize them to form a response to the user, which in the example of Figure 6.6 is "This image contains two celebrities, Kobe Bryant and Paul Pierce. They are both basketball players."

If the user continues to interact with MM-REACT, the process as described above is repeated where MM-ReAct performs planning, execution, and response generation by taking into account all observations and chat history. For instance, if the user asks "how many championship rings did the player on the left win in his career", although the answer cannot be derived from the existing observations nor chat history, the system detects the bounding boxes to decide who is on the left and the names of the players, and then plans to invoke Bing Search to find the right answer, which is "5".

6.3.2 Capabilities

Figure 6.7 shows the representative capabilities and application scenarios that MM-REACT demonstrates, including visual math and text reasoning, understanding visual-conditioned jokes/memes, spatial/coordinate understanding, visual planning and prediction, multi-image reasoning, multi-hop document understanding, open-world concept understanding, video analysis and summarization.

In addition, we show the responses generated by MM-REACT for a multi-image reasoning task in Figure 6.8. The task would be very challenging to handle simply by visual instruction tuning, as in Section 5. We refer the readers to the original MM-REACT paper for a more comprehensive demonstration of model's emergent abilities.

6.3.3 Extensibility

One favorable property of using tool chaining to build a multimodal agent is that the agent can be easily extended and enhanced from two

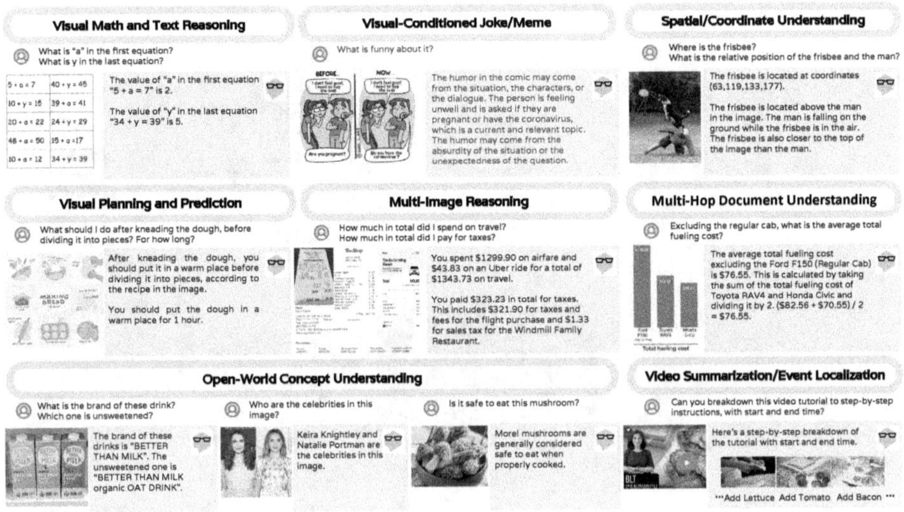

Figure 6.7: Emerging capabilities of MM-REACT for multimodal reasoning and action. Image credit: [513].

perspectives. One is to upgrade the core part of the agent, the LLM. The other is to expand the set of external tools.

Upgrading LLM. The system design of MM-REACT allows for the LLM to be upgraded to newer and more powerful ones, without the need of re-training. As shown in Figure 6.9, ChatGPT can be readily upgraded to the language-only GPT-4 model to significantly improves the performance of MM-REACT to match that of multimodal GPT-4.

Plug-and-play (adding more tools). Existing multimodal agents incorporate tools via a plug-and-play mechanism, allowing new tools to be added without training. One prominent work along this direction is HuggingGPT [396], which leverages all open-source models hosted on huggingface. Chameleon [291] incorporates not only huggingface models, but also open-source models from GitHub, Bing search API, and Python compiler. RestGPT [410] proposes a multi-level online planning framework to effectively handle the practical challenges associated with integrating LLMs with more than 100 RESTful APIs. However, it remains challenging in scaling this framework to thousands to millions of tools, as envisioned in TaskMatrix.AI [254].

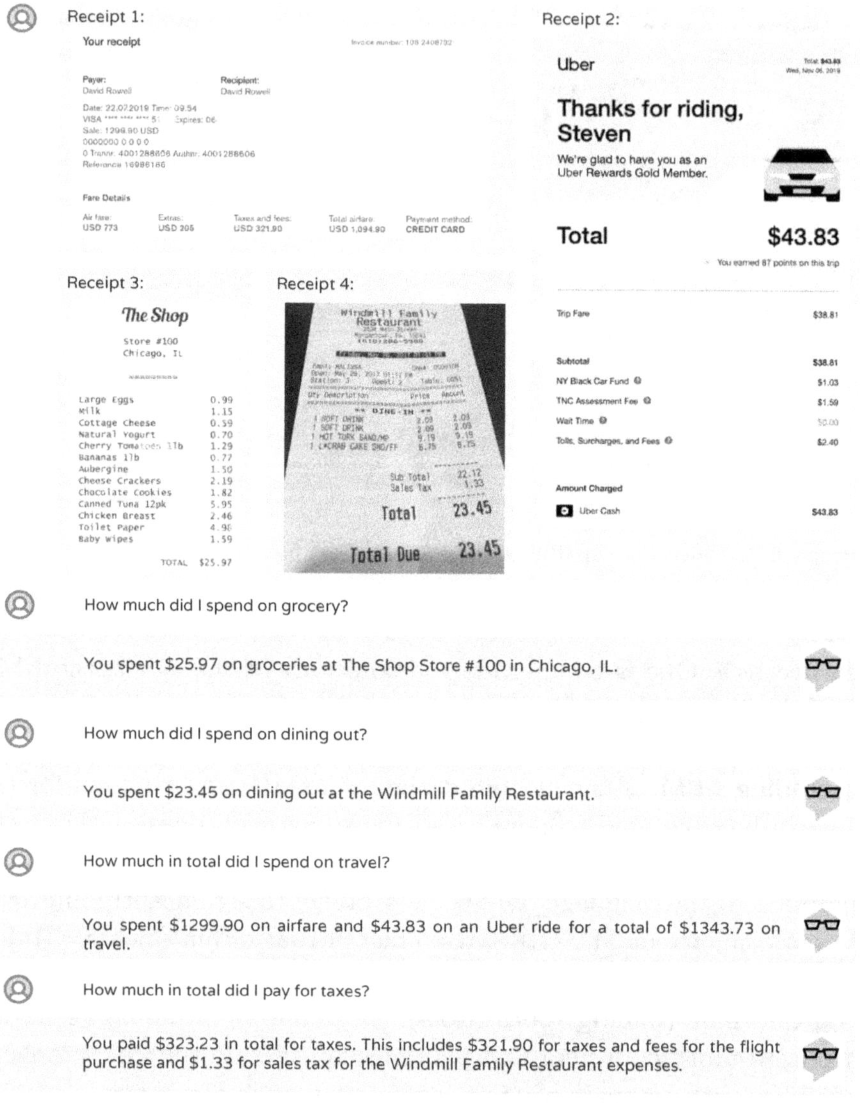

Figure 6.8: Examples of MM-ReAct's capabilities and application scenarios: **multi-image reasoning**. Image credit: [513].

Moreover, one can use SAM [206] to provide human users with diverse means, beyond text, to interact with a multimodal agent. Recall that in MM-ReAct, user intent is captured by natural language queries

I. Principe de la détection de rayonnement avec un bolomètre

Comme illustré sur la figure 1 un bolomètre est constitué d'un absorbeur qui reçoit le rayonnement que l'on désire détecter. Sa température T, supposée uniforme, est mesurée à l'aide d'un thermomètre incorporé, constitué d'un matériau conducteur dont la résistance $R(T)$ varie avec la température T : cette variation est caractérisée par le coefficient $\alpha = \frac{1}{R}\frac{dR}{dT}$. L'ensemble possède la capacité thermique C_{th}.

Un barreau, conducteur thermique, homogène, de longueur L, de section S et de conductivité thermique λ et sans échanges thermiques latéraux, relie le bolomètre à un thermostat de température T_b fixe.

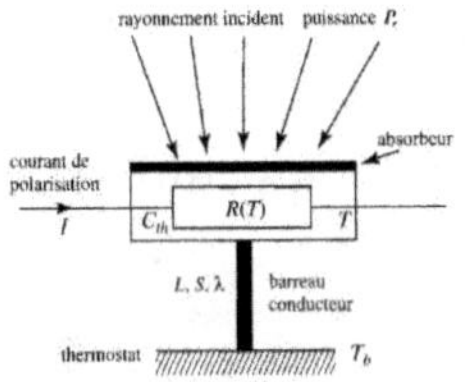

Figure 1 - Schéma d'un bolomètre

I.1. Dans cette première partie, on étudie le comportement du bolomètre en régime stationnaire, à la température de fonctionnement $T = T_0$.

I.1.a) Exprimer la température $T(x)$ en tout point du barreau conducteur repéré par son abscisse x : on prendra $T(0) = T_0$ et $T(L) = T_b$.

GPT-4 (Multimodal)

User **Answer question I.1.a. Think step-by-step.**

GPT-4 To answer question I.1.a, we need to express the temperature T(x) in every point of the conductive bar, which is characterized by the abscissa x. We are given that T(0) = T0 and T(L) = Tb.
For a stationary regime, the heat equation for the one-dimensional heat conduction in the bar is:
$d^2T/dx^2 = 0$
Integrating once, we get:
$dT/dx = C_1$
Integrating a second time:
$T(x) = C_1x + C_2$
To determine the constants C_1 and C_2, we use the boundary conditions:
T(0) = T0, which gives C_2 = T0
T(L) = Tb, which gives C_1 = (Tb - T0) / L
Finally, the expression for T(x) is:
T(x) = (Tb - T0) * (x / L) + T0
This represents a linear temperature profile along the conductive bar, with a slope of (Tb - T0) / L.

Answer question I.1.a. Think step-by-step.

MM-ReAct w/ ChatGPT

To answer question I.1.a, one would need to use the principles of thermodynamics and heat transfer to determine the temperature at any point in the conductor bar. This would likely involve solving a differential equation to find the temperature distribution along the bar. The solution would depend on the specific properties of the bar, such as its length, cross-sectional area, and thermal conductivity, as well as the temperature of the bolometer and the temperature of the thermostat. The solution would also depend on the power of the incident radiation and the current of polarization. Once the temperature distribution is found, the temperature at any point in the bar can be determined by evaluating the solution at that point.

MM-ReAct w/ GPT-4 (Language-only)

To answer question I.1.a, we'll consider the heat conduction equation along the thermal conductor bar. We're working under the assumptions of stationary mode and one-dimensional heat conduction with no lateral heat exchanges.
Start with Fourier's Law of heat conduction:
q = -k * (dT/dx)
where q is the heat flow (W/m^2), k is the thermal conductivity (W/m·K), and dT/dx is the temperature gradient along the x-axis.
In a stationary regime, the heat flow q is constant throughout the bar. Thus, we have:
-k * (dT/dx) = constant
Integrate the equation with respect to x:
∫(dT/dx) dx = ∫(constant/k) dx
After integration, we get:
T(x) = (constant/k) * x + C
where C is the integration constant.
Apply the boundary conditions to find the constants:
T(0) = To => C = To
T(L) = Tb => (constant/k) * L + To = Tb
Solve for the constant:
constant = k * (Tb - To) / L
Substitute the constant back into the T(x) equation:
T(x) = (k * (Tb - To) / L) * x / k + To
T(x) = ((Tb - To) / L) * x + To
So, the temperature T(x) at any point x along the thermal conductor bar is given by:
T(x) = ((Tb - To) / L) * x + To

Figure 6.9: Extensibility of multimodal agents: upgrading LLMs. Image credit: [513].

from the user. In InternGPT [281], connecting SAM with GPT allows users to interact with the system via clicks, scribbles, drawing bounding boxes, and so on. These interactions naturally mimic the actions of finger-pointing observed in human conversations.

6.4 Advanced Topics

In this section, we discuss several advanced topics and shed light on future research directions.

6.4.1 Compared to Training with LLMs in Section 5

We have discussed two approaches to building advanced multimodal systems based on LLMs. The one described in this section allows multimodal agents to leverage LLMs' planning abilities to select and compose various multimodal tools, while the other one described in Section 5 trains multimodal models solely based on LLMs for text generation conditioned on multimodal inputs.

These approaches have their advantages and disadvantages. On the one hand, instruction tuning, as described in Section 5, enables an end-to-end model to directly interpret rich semantics in multimodal inputs, but requires data curation and model training, hence more computationally expensive. However, limited instruction tuning data may limit the model's capabilities in such scenarios as OCR. On the other hand, one can build a multimodal agent by chaining (instead of training) LLMs with off-the-shelf models, APIs, and code interpreters, and leveraging in-context few-shot examples to teach LLMs to plan. However, as the LLM is not trained, the agent may fail to invoke correct tools given a new task. In addition, sub-optimal tools may produce noisy outputs to undermine LLM's planning or reasoning abilities, leading to suboptimal performance.

Given that these two approaches are complementary, a promising future research area is to combine the strengths of both. For example, we might replace LMM with open-source LMMs such as LLaVA [264] as a planner, and then use instruction tuning to improve LMM's planning ability, discover new abilities by composing existing tools, and improve the performance of tools if they are trainable.

6.4.2 Improving Multimodal Agents

Existing multimodal agents mainly rely on in-context few-shot examples to teach LLM to plan. This could be unreliable, leading to inaccurate tool selection and use. Below, we review the works on improving LLMs' planning performance. These works can be grouped into four categories.

Composing tools via code generation. Most existing multimodal agents use natural language to prompt an LLM for planning. Researchers [155], [419] have also explored using programming language for planning. Visual programming [155] uses in-context learning of GPT-3 [37] to generate Python-like modular programs from natural language instructions for compositional visual tasks. ViperGPT [419] instructs the GPT-3-powered Codex model [59] to generate Python codes to compose multimodal tools for a one-round query answering. However, as the codes are still generated by an LLM, the problem of inaccurate tool selection and use remains.

Improving accuracy in tool use: self-assessment. AssistGPT [130] improves the accuracy in tool selection and use via self-assessment. A stage of inspection and learning is added into the system workflow. When the round of planning and execution is finished, the system inspects the outcome, and determines whether the reasoning path of calling tools is valid, and if so, save it as an in-context example to teach the LLM to correctly select and use tools in future rounds.

Improving accuracy in tool use: instruction tuning. Instruction tuning has been exploited to improve the accuracy in tool selection and use [341], [506]. One can generate a dataset of instruction-API pairs via self-instruct to tune an LLM (*e.g.* Vicuna-7B [427]) for planning.

LMMs as tool allocators. LMMs can replace LLMs as tool allocators to power more diverse multimodel applications. An LMM can take multimodal as input. So, there is no need to unify the outputs of tools to text sequences, significantly easing the interactions between the tool allocator and multimodal tools. For instance, one could use

GPT-4V [334] to coordinate various image or video generation tools to make a short movie by providing it with a sketch story and visual examples of the main characters.

6.4.3 Diverse Applications of Multimodal Agents

Incorporating and composing tools for specific domains is an effective approach to building diverse domain-specific AI agents. [534] composes LLMs with image synthesis and understanding tools to build a data synthesis pipeline to produce diverse annotations on synthesized images. Instruct2Act [184] augments an LLM with robotic executors to generate robotic actions based on multimodal instructions. By chaining a pool of audio models with an LLM, AudioGPT [182] can understand and generate speech, music, sound, and talking head. Similarly, WavJourney [276] powers compositional audio creation of stories, rendered with speech, music, and sound. With tracking, captioning, and audio understanding models, ChatVideo [447] augments ChatGPT to understand multi-channel videos. Other application scenarios include 3D scene generation [121], [259], medical image understanding [274], [418], vision-language navigation [586], and so on.

6.4.4 Evaluation of Multimodal Agents

Using multimodal tools. It remains open how to evaluate the performance of tool selection and use of a multimodal AI agent. API-Bank [243] is one of very few efforts of systematically evaluating tool-augmented LLMs. Figure 6.10 presents an example of evaluation benchmark.

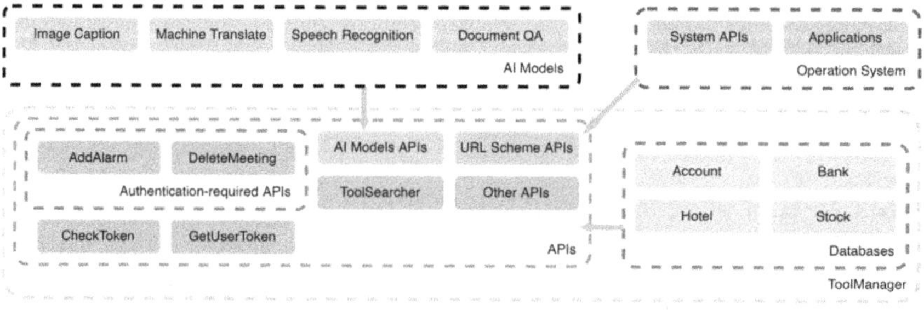

Figure 6.10: Example of evaluation benchmark focusing on toll using accuracy.

Emergent abilities. Existing VL benchmarks focus on specific model abilities, such as visual recognition [8], image description [2], [76], scene text understanding [157], [401], commonsense reasoning [545] and outside knowledge [387]. However, the emergent abilities shown in large multimodal models and agents cannot be effectively evaluated using existing benchmarks. Examples of these emergent abilities include solving math problems written on the blackboard, reasoning about events and celebrities in news images, and explaining visual jokes. Furthermore, the long, chatty outputs from these AI systems pose challenges to today's evaluation metrics. To address these challenges, researchers [123], [278], [289] have started to design comprehensive evaluation datasets to facilitate LMM evaluation. For example, MM-Vet [537] defines 6 core VL abilities and examines 16 combinations of these abilities (Figure 6.11). In addition, MM-Vet uses an LLM-based evaluator to evaluate the quality of LLM's free-form text answers across different question types and answer styles. MathVista [289] is a benchmark designed to amalgamate challenges from diverse mathematical and visual tasks.

6.4.5 Tool Creation

Consider a new task where there is no existing tool to use. Can we create a tool on the fly based on user-provided instructions? In NLP, CREATOR [356] creates tools by writing Python codes for math reasoning, rather than calling math solver APIs such as Wolfram Alpha. [40] further explores LLMs' abilities of making tools. Two LLMs are used in the study. One is used as a tool maker and the other as a tool user. They collaborate to solve complicated tasks, such as scheduling a meeting. For multimodal agents, the challenge is how to create tools that can process multimodal inputs. One may follow ViperGPT [419] to instruct LLMs to generate Python programs by leveraging pre-existent Python packages such as Open-CV. AutoML GPT [567] utilizes LLMs to automate model training. One area of future research is to develop novel multimodal deep learning tools that are tailored to effectively addressing user requests on the fly.

Figure 6.11: Example of evaluation benchmark focusing on emergent capabilities. MM-Vet focuses on the integration of different core VL capabilities, including recognition, OCR, knowledge, language generation, spatial awareness, and math. Image credit: [537].

6.4.6 Retrieval-Augmented Multimodal Agents

For many real-world applications, a substantial portion of knowledge and information that are useful for problem solving resides in external databases, because it is infeasible to encode all into the weights of any pre-trained models, particularly for long-tail concepts and fast-evolving data.

In NLP, several works augment LLMs with external data encoded with structured language and relation representations [159], [221], [344], [349]. Given input texts, such retrieved-augmented models utilize a retriever that retrieves relevant documents from an external dataset, and generates predictions given the retrieved documents.

Motivated by retrieval-augmented models in NLP, several recent works leverage visual and textual knowledge to improve vision tasks, such as image classification [283], captioning [512], question answering [67], [149], [303], [479], [510], image generation [33], [69], [397], [591], and multi-modal tasks simultaneously [518]. KAT (knowledge augmented transformer [149]) integrates implicit and explicit knowledge in an encoder-decoder architecture, and perform jointly reasoning over both knowledge sources during answer generation for open-domain VQA. RAC [283] improves long-tail classification by retrieving from a non-parametric memory that contains pre-encoded images and text. K-LITE [394] enhances the text prompts with the retrieved external knowledge that is encoded in natural language. REACT [265] retrieve from billions of the paired knowledge of image-text to improve task transfer performance for core vision problems. Among them, RA-CM3 [518] builds the first retrieval-augmented LMM with a multimodal retriever to retrieve multimodal documents, and a retrieval-augmented generator to generate both text and image. Chaining tools with LLMs shares strong similarities with the retrieval-augmented methods in that both leverage external knowledge to provide additional information for the base model to use. In the multimodal regime, an image can be used as a query to retrieve external knowledge from either a knowledge base or another pre-trained vision model.

7

Conclusions and Research Trends

Multimodal foundation models have gained a growing interest in the fields of computer vision and multimodal research. Although prevailing research topics, approaches and methodologies are quickly evolving – encompassing image self-supervised learning, language-image contrastive learning, text-to-image generation, unified vision modeling, and large language-and-vision assistants – they converge to an overarching objective: the creation of general-purpose AI agents that are capable of following human intent to perform a diverse array of vision and vision-language tasks in the wild. In this section, we summarize what have been reviewed, and present several research trends.

7.1 Summary

This monograph surveys recent advances in multimodal foundation model. We group related research in two categories, as summarized in Table 7.1.

- **Special-purpose multimodal foundation models.** There is a diverse set of tasks to tackle in computer vision. To lay the foundation for general-purpose visual assistants, we have reviewed studies

Table 7.1: Comparisons of different multimodal foundation model families covered in this monograph.

	Models	Advantages	Disadvantages
Specific-Purpose Models	Visual Understanding	Well studied and scalable solutions on image-level understanding; Emerging interests and success on region-level and pixel-level visual understanding	High training cost; No successful scalable recipe beyond the billion-image level
	Visual Generation	Well studied and scalable solutions on image-level generation; Emerging interests and success in controllable/customized image generation	High training and inference cost; Debate between diffusion and auto-regressive solutions for the best recipe; More studies are needed for video generation
General-Purpose Assistants	Unified Vision Models	Promises to unlock new emerging capabilities and scenarios	High risks in modeling and high training cost
	Training with LLM	Some new emerging capabilities and scenarios are enabled with light model training	The performance is bounded by LLM
	Chaining with LLM	Fast system development cycles with low cost as no training is involved	Low flexibility in improving system performance; No new emerging capabilities

on model pre-training, the mainstream modeling paradigm where a model is pre-trained on a large amount of task-specific data, and then transferred to a number of real-world scenarios of the same problem type in a zero- or few-shot fashion. Specifically, we have discussed two topics: (*i*) *Visual Understanding* in Section 2, where individual multimodal foundation models are developed to analyze the content of visual data in the image, region, pixel levels, respectively, and language-augmented vision models that have contributed the recent success of visual understanding tasks in the wild. (*ii*) *Visual Generation* in Section 3, where text-to-image generation models serve the foundation for image synthesis and have been successfully extended to allow users to control and customize how an image is generated. The availability and creation of large amounts of task-specific data is imperative to make these multimodal foundation models possible.

- **General-purpose assistants.** We have reviewed large numbers of emerging studies on building general-purpose assistants, which often use an unified network architecture, an unified input-output data format, and a general interface that facilitates easy interaction with humans. Motivated by the success of using LLMs such as ChatGPT/GPT-4 as a general assistant for a wide range of NLP tasks, researchers in computer vision have been exploring similar solutions for vision tasks. Based on how LLMs are used, existing works can be grouped into three categories: (*i*) *Unified Vision Models* in Section 4, where the methods of building unified LLMs for NLP tasks inspire the development of unified vision models. (*ii*) *Training with LLM* in Section 5, where visual data is curated to augment a pre-trained LLM with end-to-end training. (*iii*) *Chaining with LLM* in Section 6, where a set of vision tools (or experts) are activated by prompting a frozen LLM to complete specific vision tasks.

7.2 Case Study on Open-Source Project LLaVA

The ambition to construct a general-purpose assistant represents a monumental objective, necessitating considerable efforts and resources. In addition to proprietary models developed by individual technology companies, there is a notable presence of active open-source initiatives. These collaborative efforts, driven by the wider community, are directed towards achieving this overarching goal, reflecting a collective approach in technological advancement. LLaVA represents a pioneering endeavor in the open-source domain, aiming to construct a multimodal foundation model with capabilities extending to and surpassing those of GPT-4V. This project has been a source of inspiration for numerous initiatives focused on developing general-purpose assistants. Over the past eight months, LLaVA has garnered significant academic and community interest, evidenced by over 600 citations and in excess of 12,400 stars on GitHub.

The LLaVA project, consistently maintained by its initiating research team, has additionally stimulated a diverse array of independent and parallel endeavors. These efforts are focused on augmenting the model with a range of capabilities essential for a general-purpose assistant, thereby expanding its scope and application. Figure 7.1 shows the several development milestones of LLaVA from the initiating research team

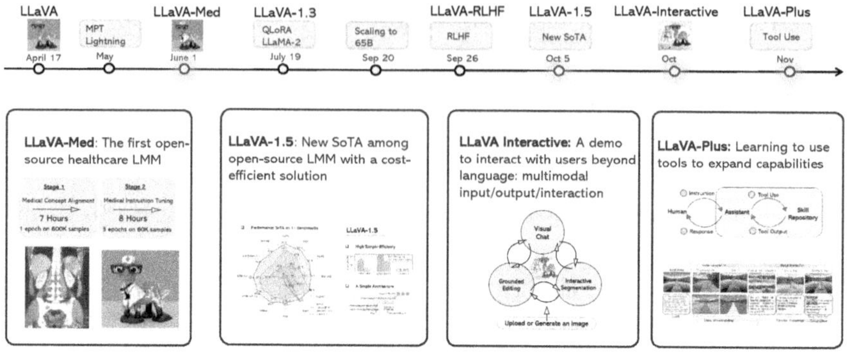

Figure 7.1: The development of LLaVA project in 2023.

in 2023. For example, LLaVA-1.5 [263] achieves new SoTA on 12 LMM benchmarks with 100× smaller compute/data. LLaVA-Interactive [66] is an all-in-one research demo for image chat, segmentation and generation/editing. It allows multimodal input, output and interaction, showing the potentials of a multimodal assistant. LLaVA-Plus [270] leverages visual instruction tuning to teach LLaVA to learning to use external tools for creating multimodal agents.

Further, we provide more examples from the community that build upon LLaVA. For visual understanding, [528] and LLaVA-Grounding [558] extends LMM with grounding and referring capability, Video-LLaVA [255] and PG-Video-LLaVA [323] expand LLaVA's capability to video understanding. ViP-LLaVA [39] extends LLaVA to understand arbitrary visual prompts, allowing users to intuitively mark images and interact with the model using natural cues like a "red bounding box" or "pointed arrow". MGIE [124] leverages LMM for instructional visual editing. Following the self-instruct visual instruction tuning data creation pipeline, LVIS-Instruct4V [448] and ShareGPT4V [58] produce higher quality data using GPT-V, boosting the multimodal model performance. μ-LLaVA [496] unifies multimodal tasks via LMM using tool use framework. LLaVA-ϕ [597] builds an efficient multimodal assistant with small language model Phi [251]. Through the collective endeavor of the community to investigate a myriad of ideas, a pathway is being forged towards the development of transparent and trustworthy general-purpose assistants.

7.3 Towards Building General-Purpose AI Agents

At the end of each section, we have discussed future trends for individual topics. This monograph is structured to demonstrate the transition from special-purpose multimodal foundation models to general-purpose multimodal AI assistants. Existing multimodal assistants, such as Flamingo [4] and multimodal GPT-4 [335], are still in the infant stage towards developing a general-purpose multimodal AI agent via foundation models. In what follows, we present several research trends.

Generalist agents with multi-modality. This aligns with the grand goal of building a single generalist agent that interacts with the world like humans through multiple channels such as language, vision, speech and actions. From this perspective, the notion of multimodal foundation models becomes somewhat indistinct on its own. Instead, it serves as a crucial component of the agent to perceive and synthesize visual signals. For example, Gato [372] and PaLM-E [108] perform a wide range of language, multimodal and control tasks with a single model, where visual perception is a crucial component in understanding the environment. It also presents challenges in the effective and scalable pre-training objectives for unified multimodal modeling.

Alignment with human intent. AI alignment research focuses on steering AI systems towards humans' goals, values, or ethical guidelines. An AI system is deemed aligned when it effectively promotes the desired goals. Though language has exhibited its generality in expressing human intent, it is not always the best option. As demonstrated in SAM [206], ControlNet [562] and GLIGEN [252], human intent can be more precisely and conveniently represented using visual prompts such as key points, bounding boxes, sketch drawing or another image, for visual understanding and generation tasks. Building foundation models that are well equipped with such a multimodal human-machine interface is a key step to unlock new use scenarios, where human intent can be well represented visually, *e.g.* the spatial arrangement of elements within a scene, and the artistic style and visual appeal of a piece of visual art.

Planning, memory, and tool use. As shown in recent studies [*e.g.* 344], [475], an LLM-powered autonomous agent can be built by using an LLM as the agent's *brain*, augmented with such key modules as planning, working memory and tool use.

Planning decomposes a large, complex task into a set of smaller, manageable sub-tasks. Ideally, an AI agent should possess the self-improvement ability such that it can evaluate the effectiveness of its previous actions and learn from errors to improve its responses to subsequent endeavors and its final outcomes. As the visual modality is a common channel to represent the state of the environment, it is important to improve models' capability of visual understanding in order to perceive salient visual features at different levels of granularity from images and videos.

Working memory (or short-term memory) is critical to in-context learning as it stores information such as user demonstrations and task descriptions that is useful for an AI agent to align its responses with user intent. Long-term memory provides the agent with the capability of storing external knowledge extracted from human-AI interactions [265]. Future foundation models need to be equipped with skills to effectively use both types of memory.

The agent also needs to learn to use external tools (or APIs) to perform the tasks that cannot be performed by the foundation model alone. For example, based on the modalities of input signals and instructions, the agent needs to use proper tools to detect user intent and complete the tasks. New tools are required to deal with new modalities in new scenarios.

The field of multimodal foundation models evolves rapidly, with new research directions and methods emerging constantly. Thus, many important research topics are not fully discussed in this monograph. We hope that we could update this monograph often enough to track the research progress.

In the near future, we expect that we would turn research ideas in computer vision into realities more quickly by following the success of LLMs, and that connecting computer vision with the broader AI community would significantly advance the development of general-purpose AI agents to make the world a better place for everyone.

Acknowledgments

This monograph is largely based on our CVPR 2023 tutorial on vision foundation models. Many people have supported us and provided valuable feedback to the writing of this monograph. We thank all the authors who have contributed to the related papers, which makes the tutorial and monograph possible. We are also grateful to Mark de Jongh, the editor from the journal of *Foundations and Trends® in Computer Graphics and Vision*, for inspiring and encouraging us to write this monograph on multimodal foundation models.

Author Contributions

Chunyuan Li initiated the project, and took lead in the writing of Sections 1, 5 and 7. Zhe Gan, Zhengyuan Yang, Jianwei Yang, Linjie Li took lead in the writing of Sections 2, 3, 4 and 6, respectively. Lijuan Wang and Jianfeng Gao provided comprehensive suggestions and edits of this work. All the authors provided project advice, and contributed to reviewing, editing and proofreading.

Chunyuan Li is currently with ByteDance/TikTok (lichunyuan24@gmail.com). Zhe Gan is currently with Apple AI/ML (zhe.gan@apple.com).

References

[1] A. Agarwal, S. Karanam, K. Joseph, A. Saxena, K. Goswami, and B. V. Srinivasan, "A-star: Test-time attention segregation and retention for text-to-image synthesis," *arXiv preprint arXiv:2306.14544*, 2023.

[2] H. Agrawal, K. Desai, Y. Wang, X. Chen, R. Jain, M. Johnson, D. Batra, D. Parikh, S. Lee, and P. Anderson, "Nocaps: Novel object captioning at scale," in *ICCV*, 2019.

[3] M. Ahn, A. Brohan, N. Brown, Y. Chebotar, O. Cortes, B. David, C. Finn, K. Gopalakrishnan, K. Hausman, A. Herzog, *et al.*, "Do as i can, not as i say: Grounding language in robotic affordances," *arXiv preprint arXiv:2204.01691*, 2022.

[4] J.-B. Alayrac, J. Donahue, P. Luc, A. Miech, I. Barr, Y. Hasson, K. Lenc, A. Mensch, K. Millican, M. Reynolds, *et al.*, "Flamingo: A visual language model for few-shot learning," *arXiv preprint arXiv:2204.14198*, 2022.

[5] M. Allahyari, S. Pouriyeh, M. Assefi, S. Safaei, E. D. Trippe, J. B. Gutierrez, and K. Kochut, "Text summarization techniques: A brief survey," *arXiv preprint arXiv:1707.02268*, 2017.

[6] E. Amrani, L. Karlinsky, and A. Bronstein, "Self-supervised classification network," in *ECCV*, 2022.

[7] P. Anderson, X. He, C. Buehler, D. Teney, M. Johnson, S. Gould, and L. Zhang, "Bottom-up and top-down attention for image captioning and visual question answering," in *CVPR*, 2018.
[8] S. Antol, A. Agrawal, J. Lu, M. Mitchell, D. Batra, C. L. Zitnick, and D. Parikh, "Vqa: Visual question answering," in *ICCV*, 2015.
[9] S. Arora, H. Khandeparkar, M. Khodak, O. Plevrakis, and N. Saunshi, "A theoretical analysis of contrastive unsupervised representation learning," *arXiv preprint arXiv:1902.09229*, 2019.
[10] M. Assran, M. Caron, I. Misra, P. Bojanowski, F. Bordes, P. Vincent, A. Joulin, M. Rabbat, and N. Ballas, "Masked siamese networks for label-efficient learning," in *ECCV*, 2022.
[11] O. Avrahami, K. Aberman, O. Fried, D. Cohen-Or, and D. Lischinski, "Break-a-scene: Extracting multiple concepts from a single image," *arXiv preprint arXiv:2305.16311*, 2023.
[12] O. Avrahami, O. Fried, and D. Lischinski, "Blended latent diffusion," *arXiv preprint arXiv:2206.02779*, 2022.
[13] O. Avrahami, T. Hayes, O. Gafni, S. Gupta, Y. Taigman, D. Parikh, D. Lischinski, O. Fried, and X. Yin, "Spatext: Spatio-textual representation for controllable image generation," in *Proceedings of the IEEE/CVF Conference on Computer Vision and Pattern Recognition*, pp. 18 370–18 380, 2023.
[14] O. Avrahami, D. Lischinski, and O. Fried, "Blended diffusion for text-driven editing of natural images," in *Proceedings of the IEEE/CVF Conference on Computer Vision and Pattern Recognition*, pp. 18 208–18 218, 2022.
[15] A. Awadalla, I. Gao, J. Gardner, J. Hessel, Y. Hanafy, W. Zhu, K. Marathe, Y. Bitton, S. Gadre, J. Jitsev, S. Kornblith, P. W. Koh, G. Ilharco, M. Wortsman, and L. Schmidt, *Openflamingo*, version v0.1.1, Mar. 2023. DOI: 10.5281/zenodo.7733589.
[16] M. Awais, M. Naseer, S. Khan, R. M. Anwer, H. Cholakkal, M. Shah, M.-H. Yang, and F. S. Khan, "Foundational models defining a new era in vision: A survey and outlook," *arXiv preprint arXiv:2307.13721*, 2023.
[17] P. Bachman, R. D. Hjelm, and W. Buchwalter, "Learning representations by maximizing mutual information across views," *NeurIPS*, 2019.

[18] A. Baevski, W.-N. Hsu, Q. Xu, A. Babu, J. Gu, and M. Auli, "Data2vec: A general framework for self-supervised learning in speech, vision and language," in *ICML*, 2022.

[19] D. Bahdanau, K. Cho, and Y. Bengio, "Neural machine translation by jointly learning to align and translate," in *ICLR*, 2015.

[20] J. Bai, S. Bai, S. Yang, S. Wang, S. Tan, P. Wang, J. Lin, C. Zhou, and J. Zhou, "Qwen-vl: A frontier large vision-language model with versatile abilities," *arXiv preprint arXiv:2308.12966*, 2023.

[21] S. Bai, S. Yang, J. Bai, P. Wang, X. Zhang, J. Lin, X. Wang, C. Zhou, and J. Zhou, *Touchstone: Evaluating vision-language models by language models*, 2023. URL: https://arxiv.org/abs/2308.16890.

[22] Y. Balaji, S. Nah, X. Huang, A. Vahdat, J. Song, K. Kreis, M. Aittala, T. Aila, S. Laine, B. Catanzaro, *et al.*, "Ediffi: Text-to-image diffusion models with an ensemble of expert denoisers," *arXiv preprint arXiv:2211.01324*, 2022.

[23] I. Balažević, D. Steiner, N. Parthasarathy, R. Arandjelović, and O. J. Hénaff, "Towards in-context scene understanding," *arXiv preprint arXiv:2306.01667*, 2023.

[24] A. Bansal, K. Sikka, G. Sharma, R. Chellappa, and A. Divakaran, "Zero-shot object detection," in *Proceedings of the European conference on computer vision (ECCV)*, pp. 384–400, 2018.

[25] A. Bansal, H.-M. Chu, A. Schwarzschild, S. Sengupta, M. Goldblum, J. Geiping, and T. Goldstein, "Universal guidance for diffusion models," in *Proceedings of the IEEE/CVF Conference on Computer Vision and Pattern Recognition*, pp. 843–852, 2023.

[26] H. Bao, L. Dong, and F. Wei, "BEiT: Bert pre-training of image transformers," in *ICLR*, 2022.

[27] A. Bar, Y. Gandelsman, T. Darrell, A. Globerson, and A. Efros, "Visual prompting via image inpainting," *Advances in Neural Information Processing Systems*, vol. 35, 2022, pp. 25 005–25 017.

[28] A. Bardes, J. Ponce, and Y. LeCun, "Vicreg: Variance-invariance-covariance regularization for self-supervised learning," *arXiv preprint arXiv:2105.04906*, 2021.

[29] J. Berant, A. Chou, R. Frostig, and P. Liang, "Semantic parsing on freebase from question-answer pairs," in *Proceedings of the 2013 conference on empirical methods in natural language processing*, pp. 1533–1544, 2013.

[30] Y. Bitton, H. Bansal, J. Hessel, R. Shao, W. Zhu, A. Awadalla, J. Gardner, R. Taori, and L. Schimdt, *Visit-bench: A benchmark for vision-language instruction following inspired by real-world use*, 2023. arXiv: 2308.06595.

[31] K. Black, M. Janner, Y. Du, I. Kostrikov, and S. Levine, "Training diffusion models with reinforcement learning," *arXiv preprint arXiv:2305.13301*, 2023.

[32] A. Blattmann, T. Dockhorn, S. Kulal, D. Mendelevitch, M. Kilian, D. Lorenz, Y. Levi, Z. English, V. Voleti, A. Letts, V. Jampani, and R. Rombach, "Stable video diffusion: Scaling latent video diffusion models to large datasets," 2023.

[33] A. Blattmann, R. Rombach, K. Oktay, and B. Ommer, "Retrieval-augmented diffusion models," *arXiv preprint arXiv:2204.11824*, 2022.

[34] D. Bolya, C. Zhou, F. Xiao, and Y. J. Lee, "Yolact: Real-time instance segmentation," in *Proceedings of the IEEE/CVF International Conference on Computer Vision*, pp. 9157–9166, 2019.

[35] R. Bommasani, D. A. Hudson, E. Adeli, R. Altman, S. Arora, S. von Arx, M. S. Bernstein, J. Bohg, A. Bosselut, E. Brunskill, *et al.*, "On the opportunities and risks of foundation models," *arXiv preprint arXiv:2108.07258*, 2021.

[36] T. Brooks, A. Holynski, and A. A. Efros, "Instructpix2pix: Learning to follow image editing instructions," in *Proceedings of the IEEE/CVF Conference on Computer Vision and Pattern Recognition*, pp. 18 392–18 402, 2023.

[37] T. B. Brown, B. Mann, N. Ryder, M. Subbiah, J. Kaplan, P. Dhariwal, A. Neelakantan, P. Shyam, G. Sastry, A. Askell, *et al.*, "Language models are few-shot learners," in *NeuIPS*, 2020.

[38] M. Byeon, B. Park, H. Kim, S. Lee, W. Baek, and S. Kim, *Coyo-700m: Image-text pair dataset*, 2022. URL: https://github.com/kakaobrain/coyo-dataset.

[39] M. Cai, H. Liu, S. K. Mustikovela, G. P. Meyer, Y. Chai, D. Park, and Y. J. Lee, "Making large multimodal models understand arbitrary visual prompts," in *arXiv:2312.00784*, 2023.

[40] T. Cai, X. Wang, T. Ma, X. Chen, and D. Zhou, "Large language models as tool makers," *arXiv preprint arXiv:2305.17126*, 2023.

[41] Z. Cai, G. Kwon, A. Ravichandran, E. Bas, Z. Tu, R. Bhotika, and S. Soatto, "X-detr: A versatile architecture for instance-wise vision-language tasks," in *ECCV*, 2022.

[42] L. Cao, B. Zhang, C. Chen, Y. Yang, X. Du, W. Zhang, Z. Lu, and Y. Zheng, "Less is more: Removing text-regions improves clip training efficiency and robustness," *arXiv preprint arXiv:2305.05095*, 2023.

[43] N. Carion, F. Massa, G. Synnaeve, N. Usunier, A. Kirillov, and S. Zagoruyko, "End-to-end object detection with transformers," in *ECCV*, 2020.

[44] M. Caron, P. Bojanowski, A. Joulin, and M. Douze, "Deep clustering for unsupervised learning of visual features," in *ECCV*, 2018.

[45] M. Caron, I. Misra, J. Mairal, P. Goyal, P. Bojanowski, and A. Joulin, "Unsupervised learning of visual features by contrasting cluster assignments," *NeurIPS*, 2020.

[46] M. Caron, H. Touvron, I. Misra, H. Jégou, J. Mairal, P. Bojanowski, and A. Joulin, "Emerging properties in self-supervised vision transformers," in *ICCV*, 2021.

[47] L. Castrejon, K. Kundu, R. Urtasun, and S. Fidler, "Annotating object instances with a polygon-rnn," in *Proceedings of the IEEE conference on computer vision and pattern recognition*, pp. 5230–5238, 2017.

[48] H. Chang, H. Zhang, J. Barber, A. Maschinot, J. Lezama, L. Jiang, M.-H. Yang, K. Murphy, W. T. Freeman, M. Rubinstein, *et al.*, "Muse: Text-to-image generation via masked generative transformers," *arXiv preprint arXiv:2301.00704*, 2023.

[49] H. Chang, H. Zhang, L. Jiang, C. Liu, and W. T. Freeman, "Maskgit: Masked generative image transformer," in *Proceedings of the IEEE/CVF Conference on Computer Vision and Pattern Recognition*, pp. 11 315–11 325, 2022.

[50] S. Changpinyo, P. Sharma, N. Ding, and R. Soricut, "Conceptual 12m: Pushing web-scale image-text pre-training to recognize long-tail visual concepts," in *CVPR*, 2021.

[51] H. Chefer, Y. Alaluf, Y. Vinker, L. Wolf, and D. Cohen-Or, "Attend-and-excite: Attention-based semantic guidance for text-to-image diffusion models," *arXiv preprint arXiv:2301.13826*, 2023.

[52] C. Chen, B. Zhang, L. Cao, J. Shen, T. Gunter, A. M. Jose, A. Toshev, J. Shlens, R. Pang, and Y. Yang, "Stair: Learning sparse text and image representation in grounded tokens," *arXiv preprint arXiv:2301.13081*, 2023.

[53] D. Chen, J. Liu, W. Dai, and B. Wang, "Visual instruction tuning with polite flamingo," *arXiv preprint arXiv:2307.01003*, 2023.

[54] F. Chen, D. Zhang, M. Han, X. Chen, J. Shi, S. Xu, and B. Xu, "Vlp: A survey on vision-language pre-training," *arXiv preprint arXiv:2202.09061*, 2022.

[55] K. Chen, Z. Zhang, W. Zeng, R. Zhang, F. Zhu, and R. Zhao, "Shikra: Unleashing multimodal llm's referential dialogue magic," *arXiv preprint arXiv:2306.15195*, 2023.

[56] L. Chen, M. Zhai, J. He, and G. Mori, "Object grounding via iterative context reasoning," in *Proceedings of the IEEE/CVF International Conference on Computer Vision Workshops*, pp. 0–0, 2019.

[57] L.-C. Chen, G. Papandreou, F. Schroff, and H. Adam, "Rethinking atrous convolution for semantic image segmentation," *arXiv preprint arXiv:1706.05587*, 2017.

[58] L. Chen, J. Li, X. Dong, P. Zhang, C. He, J. Wang, F. Zhao, and D. Lin, "Sharegpt4v: Improving large multi-modal models with better captions," *arXiv preprint arXiv:2311.12793*, 2023.

[59] M. Chen, J. Tworek, H. Jun, Q. Yuan, H. P. d. O. Pinto, J. Kaplan, H. Edwards, Y. Burda, N. Joseph, G. Brockman, *et al.*, "Evaluating large language models trained on code," *arXiv preprint arXiv:2107.03374*, 2021.

[60] M. Chen, I. Laina, and A. Vedaldi, "Training-free layout control with cross-attention guidance," *arXiv preprint arXiv:2304.03373*, 2023.

[61] Q. Chen, X. Chen, G. Zeng, and J. Wang, "Group detr: Fast training convergence with decoupled one-to-many label assignment," *arXiv preprint arXiv:2207.13085*, 2022.

[62] T. Chen, S. Kornblith, M. Norouzi, and G. Hinton, "A simple framework for contrastive learning of visual representations," in *ICML*, 2020.

[63] T. Chen, S. Kornblith, K. Swersky, M. Norouzi, and G. E. Hinton, "Big self-supervised models are strong semi-supervised learners," *NeurIPS*, 2020.

[64] T. Chen, S. Saxena, L. Li, D. J. Fleet, and G. Hinton, "Pix2seq: A language modeling framework for object detection," in *ICLR*, 2022.

[65] T. Chen, S. Saxena, L. Li, T.-Y. Lin, D. J. Fleet, and G. Hinton, "A unified sequence interface for vision tasks," *arXiv preprint arXiv:2206.07669*, 2022.

[66] W.-G. Chen, I. Spiridonova, J. Yang, J. Gao, and C. Li, "Llava-interactive: An all-in-one demo for image chat, segmentation, generation and editing," 2023.

[67] W. Chen, H. Hu, X. Chen, P. Verga, and W. W. Cohen, "Murag: Multimodal retrieval-augmented generator for open question answering over images and text," *arXiv preprint arXiv:2210.02928*, 2022.

[68] W. Chen, H. Hu, Y. Li, N. Rui, X. Jia, M.-W. Chang, and W. W. Cohen, "Subject-driven text-to-image generation via apprenticeship learning," *arXiv preprint arXiv:2304.00186*, 2023.

[69] W. Chen, H. Hu, C. Saharia, and W. W. Cohen, "Re-imagen: Retrieval-augmented text-to-image generator," *arXiv preprint arXiv:2209.14491*, 2022.

[70] X. Chen, J. Djolonga, P. Padlewski, B. Mustafa, S. Changpinyo, J. Wu, C. R. Ruiz, S. Goodman, X. Wang, Y. Tay, *et al.*, "Pali-x: On scaling up a multilingual vision and language model," *arXiv preprint arXiv:2305.18565*, 2023.

[71] X. Chen, X. Wang, S. Changpinyo, A. Piergiovanni, P. Padlewski, D. Salz, S. Goodman, A. Grycner, B. Mustafa, L. Beyer, *et al.*, "Pali: A jointly-scaled multilingual language-image model," *arXiv preprint arXiv:2209.06794*, 2022.

[72] X. Chen, Z. Zhao, F. Yu, Y. Zhang, and M. Duan, "Conditional diffusion for interactive segmentation," in *Proceedings of the IEEE/CVF International Conference on Computer Vision*, pp. 7345–7354, 2021.

[73] X. Chen, Z. Zhao, Y. Zhang, M. Duan, D. Qi, and H. Zhao, "Focalclick: Towards practical interactive image segmentation," in *Proceedings of the IEEE/CVF Conference on Computer Vision and Pattern Recognition*, pp. 1300–1309, 2022.

[74] X. Chen, M. Ding, X. Wang, Y. Xin, S. Mo, Y. Wang, S. Han, P. Luo, G. Zeng, and J. Wang, "Context autoencoder for self-supervised representation learning," *arXiv preprint arXiv:2202.03026*, 2022.

[75] X. Chen, H. Fan, R. Girshick, and K. He, "Improved baselines with momentum contrastive learning," *arXiv preprint arXiv:2003.04297*, 2020.

[76] X. Chen, H. Fang, T. Lin, R. Vedantam, S. Gupta, P. Dollár, and C. L. Zitnick, "Microsoft COCO captions: Data collection and evaluation server," *arXiv preprint arXiv:1504.00325*, 2015.

[77] X. Chen and K. He, "Exploring simple siamese representation learning," in *CVPR*, 2021.

[78] X. Chen, S. Xie, and K. He, "An empirical study of training self-supervised vision transformers," in *ICCV*, 2021.

[79] Y.-C. Chen, L. Li, L. Yu, A. El Kholy, F. Ahmed, Z. Gan, Y. Cheng, and J. Liu, "UNITER: Universal image-text representation learning," in *ECCV*, 2020.

[80] Z. Chen, Y. Duan, W. Wang, J. He, T. Lu, J. Dai, and Y. Qiao, "Vision transformer adapter for dense predictions," *arXiv preprint arXiv:2205.08534*, 2022.

[81] B. Cheng, I. Misra, A. G. Schwing, A. Kirillov, and R. Girdhar, "Masked-attention mask transformer for universal image segmentation," in *Proceedings of the IEEE/CVF Conference on Computer Vision and Pattern Recognition*, pp. 1290–1299, 2022.

[82] M. Cherti, R. Beaumont, R. Wightman, M. Wortsman, G. Ilharco, C. Gordon, C. Schuhmann, L. Schmidt, and J. Jitsev, "Reproducible scaling laws for contrastive language-image learning," in *CVPR*, 2023.

[83] J. Cho, J. Lei, H. Tan, and M. Bansal, "Unifying vision-and-language tasks via text generation," in *ICML*, 2021.

[84] J. Cho, L. Li, Z. Yang, Z. Gan, L. Wang, and M. Bansal, "Diagnostic benchmark and iterative inpainting for layout-guided image generation," *arXiv preprint arXiv:2304.06671*, 2023.

[85] A. Chowdhery, S. Narang, J. Devlin, M. Bosma, G. Mishra, A. Roberts, P. Barham, H. W. Chung, C. Sutton, S. Gehrmann, *et al.*, "Palm: Scaling language modeling with pathways," *arXiv preprint arXiv:2204.02311*, 2022.

[86] T. Computer, *Redpajama-data: An open source recipe to reproduce llama training dataset*, 2023. URL: https://github.com/togethercomputer/RedPajama-Data.

[87] M. Crawshaw, "Multi-task learning with deep neural networks: A survey," *arXiv preprint arXiv:2009.09796*, 2020.

[88] A. Creswell, T. White, V. Dumoulin, K. Arulkumaran, B. Sengupta, and A. A. Bharath, "Generative adversarial networks: An overview," *IEEE signal processing magazine*, vol. 35, no. 1, 2018, pp. 53–65.

[89] H. Dai, C. Ma, Z. Liu, Y. Li, P. Shu, X. Wei, L. Zhao, Z. Wu, D. Zhu, W. Liu, *et al.*, "Samaug: Point prompt augmentation for segment anything model," *arXiv preprint arXiv:2307.01187*, 2023.

[90] W. Dai, J. Li, D. Li, A. M. H. Tiong, J. Zhao, W. Wang, B. Li, P. Fung, and S. Hoi, "Instructblip: Towards general-purpose vision-language models with instruction tuning," *arXiv preprint arXiv:2305.06500*, 2023.

[91] X. Dai, Y. Chen, B. Xiao, D. Chen, M. Liu, L. Yuan, and L. Zhang, "Dynamic head: Unifying object detection heads with attentions," in *Proceedings of the IEEE/CVF conference on computer vision and pattern recognition*, pp. 7373–7382, 2021.

[92] R. Datta, D. Joshi, J. Li, and J. Z. Wang, "Image retrieval: Ideas, influences, and trends of the new age," *ACM Computing Surveys (Csur)*, vol. 40, no. 2, 2008, pp. 1–60.

[93] C. Deng, Q. Wu, Q. Wu, F. Hu, F. Lyu, and M. Tan, "Visual grounding via accumulated attention," in *CVPR*, 2018.

[94] J. Deng, W. Dong, R. Socher, L.-J. Li, K. Li, and L. Fei-Fei, "Imagenet: A large-scale hierarchical image database," in *CVPR*, 2009.

[95] K. Desai and J. Johnson, "Virtex: Learning visual representations from textual annotations," in *CVPR*, 2021.

[96] K. Desai, G. Kaul, Z. Aysola, and J. Johnson, "Redcaps: Web-curated image-text data created by the people, for the people," in *NeurIPS, Track on Datasets and Benchmarks*, 2021.

[97] T. Dettmers, A. Pagnoni, A. Holtzman, and L. Zettlemoyer, "Qlora: Efficient finetuning of quantized llms," *arXiv preprint arXiv:2305.14314*, 2023.

[98] J. Devlin, M.-W. Chang, K. Lee, and K. Toutanova, "Bert: Pre-training of deep bidirectional transformers for language understanding," in *NAACL*, 2019.

[99] P. Dhariwal and A. Nichol, "Diffusion models beat gans on image synthesis," in *NeurIPS*, 2021.

[100] J. Ding, N. Xue, G.-S. Xia, and D. Dai, *Decoupling zero-shot semantic segmentation*, 2022. arXiv: 2112.07910 `[cs.CV]`.

[101] Z. Ding, J. Wang, and Z. Tu, "Open-vocabulary panoptic segmentation with maskclip," *arXiv preprint arXiv:2208.08984*, 2022.

[102] B. Dong, F. Zeng, T. Wang, X. Zhang, and Y. Wei, "Solq: Segmenting objects by learning queries," *Advances in Neural Information Processing Systems*, vol. 34, 2021, pp. 21 898–21 909.

[103] X. Dong, J. Bao, T. Zhang, D. Chen, W. Zhang, L. Yuan, D. Chen, F. Wen, and N. Yu, "Bootstrapped masked autoencoders for vision bert pretraining," in *ECCV*, 2022.

[104] X. Dong, J. Bao, T. Zhang, D. Chen, W. Zhang, L. Yuan, D. Chen, F. Wen, N. Yu, and B. Guo, "Peco: Perceptual codebook for bert pre-training of vision transformers," in *AAAI*, 2023.

[105] A. Dosovitskiy, L. Beyer, A. Kolesnikov, D. Weissenborn, X. Zhai, T. Unterthiner, M. Dehghani, M. Minderer, G. Heigold, S. Gelly, *et al.*, "An image is worth 16x16 words: Transformers for image recognition at scale," in *ICLR*, 2021.

[106] Z.-Y. Dou, A. Kamath, Z. Gan, P. Zhang, J. Wang, L. Li, Z. Liu, C. Liu, Y. LeCun, N. Peng, J. Gao, and L. Wang, "Coarse-to-fine vision-language pre-training with fusion in the backbone," in *NeurIPS*, 2022.

[107] Z.-Y. Dou, Y. Xu, Z. Gan, J. Wang, S. Wang, L. Wang, C. Zhu, Z. Liu, M. Zeng, *et al.*, "An empirical study of training end-to-end vision-and-language transformers," in *CVPR*, 2022.

[108] D. Driess, F. Xia, M. S. Sajjadi, C. Lynch, A. Chowdhery, B. Ichter, A. Wahid, J. Tompson, Q. Vuong, T. Yu, *et al.*, "PaLM-E: An embodied multimodal language model," *arXiv preprint arXiv:2303.03378*, 2023.

[109] Y. Du, Z. Liu, J. Li, and W. X. Zhao, "A survey of vision-language pre-trained models," in *IJCAI survey track*, 2022.

[110] O. Elharrouss, N. Almaadeed, S. Al-Maadeed, and Y. Akbari, "Image inpainting: A review," *Neural Processing Letters*, vol. 51, 2020, pp. 2007–2028.

[111] A. Ermolov, A. Siarohin, E. Sangineto, and N. Sebe, "Whitening for self-supervised representation learning," in *ICML*, 2021.

[112] P. Esser, R. Rombach, and B. Ommer, "Taming transformers for high-resolution image synthesis," in *CVPR*, 2021.

[113] M. Everingham and J. Winn, "The pascal visual object classes challenge 2012 (voc2012) development kit," *Pattern Analysis, Statistical Modelling and Computational Learning, Tech. Rep*, vol. 8, no. 5, 2011.

[114] L. Fan, D. Krishnan, P. Isola, D. Katabi, and Y. Tian, "Improving clip training with language rewrites," *arXiv preprint arXiv:2305.20088*, 2023.

[115] Y. Fan, O. Watkins, Y. Du, H. Liu, M. Ryu, C. Boutilier, P. Abbeel, M. Ghavamzadeh, K. Lee, and K. Lee, "Dpok: Reinforcement learning for fine-tuning text-to-image diffusion models," *arXiv preprint arXiv:2305.16381*, 2023.

[116] Y. Fang, W. Wang, B. Xie, Q. Sun, L. Wu, X. Wang, T. Huang, X. Wang, and Y. Cao, "Eva: Exploring the limits of masked visual representation learning at scale," in *CVPR*, 2023.

[117] Z. Fang, J. Wang, X. Hu, L. Liang, Z. Gan, L. Wang, Y. Yang, and Z. Liu, "Injecting semantic concepts into end-to-end image captioning," in *CVPR*, 2022.

[118] C. Feichtenhofer, Y. Li, K. He, *et al.*, "Masked autoencoders as spatiotemporal learners," *NeurIPS*, 2022.

[119] C. Feng, Y. Zhong, Z. Jie, X. Chu, H. Ren, X. Wei, W. Xie, and L. Ma, "Promptdet: Towards open-vocabulary detection using uncurated images," in *European Conference on Computer Vision*, Springer, pp. 701–717, 2022.

[120] W. Feng, X. He, T.-J. Fu, V. Jampani, A. R. Akula, P. Narayana, S. Basu, X. E. Wang, and W. Y. Wang, "Training-free structured diffusion guidance for compositional text-to-image synthesis," in *The Eleventh International Conference on Learning Representations*, 2022.

[121] W. Feng, W. Zhu, T.-j. Fu, V. Jampani, A. Akula, X. He, S. Basu, X. E. Wang, and W. Y. Wang, "Layoutgpt: Compositional visual planning and generation with large language models," *arXiv preprint arXiv:2305.15393*, 2023.

[122] A. Frome, G. S. Corrado, J. Shlens, S. Bengio, J. Dean, M. Ranzato, and T. Mikolov, "Devise: A deep visual-semantic embedding model," in *NeurIPS*, 2013.

[123] C. Fu, P. Chen, Y. Shen, Y. Qin, M. Zhang, X. Lin, Z. Qiu, W. Lin, J. Yang, X. Zheng, *et al.*, "Mme: A comprehensive evaluation benchmark for multimodal large language models," *arXiv preprint arXiv:2306.13394*, 2023.

[124] T.-J. Fu, W. Hu, X. Du, W. Y. Wang, Y. Yang, and Z. Gan, "Guiding instruction-based image editing via multimodal large language models," Sep. 2023.

[125] S. Y. Gadre, G. Ilharco, A. Fang, J. Hayase, G. Smyrnis, T. Nguyen, R. Marten, M. Wortsman, D. Ghosh, J. Zhang, *et al.*, "Datacomp: In search of the next generation of multimodal datasets," *arXiv preprint arXiv:2304.14108*, 2023.

[126] O. Gafni, A. Polyak, O. Ashual, S. Sheynin, D. Parikh, and Y. Taigman, "Make-a-scene: Scene-based text-to-image generation with human priors," *arXiv preprint arXiv:2203.13131*, 2022.

[127] R. Gal, Y. Alaluf, Y. Atzmon, O. Patashnik, A. H. Bermano, G. Chechik, and D. Cohen-Or, "An image is worth one word: Personalizing text-to-image generation using textual inversion," *arXiv preprint arXiv:2208.01618*, 2022.

[128] Z. Gan, Y.-C. Chen, L. Li, C. Zhu, Y. Cheng, and J. Liu, "Large-scale adversarial training for vision-and-language representation learning," in *NeurIPS*, 2020.

[129] Z. Gan, L. Li, C. Li, L. Wang, Z. Liu, J. Gao, *et al.*, "Vision-language pre-training: Basics, recent advances, and future trends," *Foundations and Trends® in Computer Graphics and Vision*, 2022.

[130] D. Gao, L. Ji, L. Zhou, K. Q. Lin, J. Chen, Z. Fan, and M. Z. Shou, "Assistgpt: A general multi-modal assistant that can plan, execute, inspect, and learn," *arXiv preprint arXiv:2306.08640*, 2023.

[131] P. Gao, J. Han, R. Zhang, Z. Lin, S. Geng, A. Zhou, W. Zhang, P. Lu, C. He, X. Yue, *et al.*, "Llama-adapter v2: Parameter-efficient visual instruction model," *arXiv preprint arXiv:2304.15010*, 2023.

[132] P. Gao, T. Ma, H. Li, Z. Lin, J. Dai, and Y. Qiao, "Convmae: Masked convolution meets masked autoencoders," *arXiv preprint arXiv:2205.03892*, 2022.

[133] Y. Ge, Y. Ge, Z. Zeng, X. Wang, and Y. Shan, "Planting a seed of vision in large language model," *arXiv preprint arXiv:2307.08041*, 2023.

[134] *Gen-2*, https://research.runwayml.com/gen2.

[135] X. Geng and H. Liu, *Openllama: An open reproduction of llama*, May 2023. URL: https://github.com/openlm-research/open_llama.

[136] Z. Geng, B. Yang, T. Hang, C. Li, S. Gu, T. Zhang, J. Bao, Z. Zhang, H. Hu, D. Chen, *et al.*, "Instructdiffusion: A generalist modeling interface for vision tasks," *arXiv preprint arXiv:2309.03895*, 2023.

[137] G. Ghiasi, X. Gu, Y. Cui, and T.-Y. Lin, "Open-vocabulary image segmentation," in *ECCV*, 2022.

[138] G. Ghiasi, X. Gu, Y. Cui, and T.-Y. Lin, "Scaling open-vocabulary image segmentation with image-level labels," in *European Conference on Computer Vision*, Springer, pp. 540–557, 2022.

[139] R. Girdhar, A. El-Nouby, Z. Liu, M. Singh, K. V. Alwala, A. Joulin, and I. Misra, "Imagebind: One embedding space to bind them all," in *CVPR*, 2023.

[140] R. Girshick, "Fast r-cnn," in *ICCV*, 2015.

[141] R. Girshick, J. Donahue, T. Darrell, and J. Malik, "Region-based convolutional networks for accurate object detection and segmentation," *IEEE transactions on pattern analysis and machine intelligence*, vol. 38, no. 1, 2015, pp. 142–158.

[142] T. Gong, C. Lyu, S. Zhang, Y. Wang, M. Zheng, Q. Zhao, K. Liu, W. Zhang, P. Luo, and K. Chen, "Multimodal-gpt: A vision and language model for dialogue with humans," *arXiv preprint arXiv:2305.04790*, 2023.

[143] I. Goodfellow, J. Pouget-Abadie, M. Mirza, B. Xu, D. Warde-Farley, S. Ozair, A. Courville, and Y. Bengio, "Generative adversarial networks," *Communications of the ACM*, 2020.

[144] J.-B. Grill, F. Strub, F. Altché, C. Tallec, P. Richemond, E. Buchatskaya, C. Doersch, B. Avila Pires, Z. Guo, M. Gheshlaghi Azar, *et al.*, "Bootstrap your own latent-a new approach to self-supervised learning," *NeurIPS*, 2020.

[145] X. Gu, Y. Cui, J. Huang, A. Rashwan, X. Yang, X. Zhou, G. Ghiasi, W. Kuo, H. Chen, L.-C. Chen, *et al.*, "Dataseg: Taming a universal multi-dataset multi-task segmentation model," *arXiv preprint arXiv:2306.01736*, 2023.

[146] X. Gu, T.-Y. Lin, W. Kuo, and Y. Cui, "Open-vocabulary object detection via vision and language knowledge distillation," *arXiv preprint arXiv:2104.13921*, 2021.

[147] X. Gu, T.-Y. Lin, W. Kuo, and Y. Cui, "Open-vocabulary object detection via vision and language knowledge distillation," in *ICLR*, 2022.

[148] A. Gudibande, E. Wallace, C. Snell, X. Geng, H. Liu, P. Abbeel, S. Levine, and D. Song, "The false promise of imitating proprietary llms," *arXiv preprint arXiv:2305.15717*, 2023.

[149] L. Gui, B. Wang, Q. Huang, A. Hauptmann, Y. Bisk, and J. Gao, "Kat: A knowledge augmented transformer for vision-and-language," in *NAACL*, 2022.

[150] A. Gunjal, J. Yin, and E. Bas, "Detecting and preventing hallucinations in large vision language models," *arXiv preprint arXiv:2308.06394*, 2023.

[151] T. Gupta, A. Kamath, A. Kembhavi, and D. Hoiem, "Towards general purpose vision systems: An end-to-end task-agnostic vision-language architecture," in *CVPR*, 2022.

[152] T. Gupta, A. Kamath, A. Kembhavi, and D. Hoiem, "Towards general purpose vision systems: An end-to-end task-agnostic vision-language architecture," in *Proceedings of the IEEE/CVF Conference on Computer Vision and Pattern Recognition (CVPR)*, pp. 16 399–16 409, Jun. 2022.

[153] T. Gupta and A. Kembhavi, "Visual programming: Compositional visual reasoning without training," *arXiv preprint arXiv:2211.11559*, 2022.

[154] T. Gupta and A. Kembhavi, "Visual programming: Compositional visual reasoning without training," *ArXiv*, vol. abs/2211.11559, 2022.

[155] T. Gupta and A. Kembhavi, "Visual programming: Compositional visual reasoning without training," in *CVPR*, 2023.

[156] T. Gupta, R. Marten, A. Kembhavi, and D. Hoiem, "Grit: General robust image task benchmark," *arXiv preprint arXiv:2204.13653*, 2022.

[157] D. Gurari, Q. Li, A. J. Stangl, A. Guo, C. Lin, K. Grauman, J. Luo, and J. P. Bigham, "Vizwiz grand challenge: Answering visual questions from blind people," in *CVPR*, 2018.

[158] M. Gutmann and A. Hyvärinen, "Noise-contrastive estimation: A new estimation principle for unnormalized statistical models," in *AISTATS*, 2010.

[159] K. Guu, K. Lee, Z. Tung, P. Pasupat, and M.-W. Chang, "Realm: Retrieval-augmented language model pre-training," *arXiv preprint arXiv:2002.08909*, 2020.

[160] A. M. Hafiz and G. M. Bhat, "A survey on instance segmentation: State of the art," *International journal of multimedia information retrieval*, 2020.

[161] A. W. Harley, Z. Fang, and K. Fragkiadaki, "Particle video revisited: Tracking through occlusions using point trajectories," in *European Conference on Computer Vision*, Springer, pp. 59–75, 2022.

[162] K. He, X. Chen, S. Xie, Y. Li, P. Dollár, and R. Girshick, "Masked autoencoders are scalable vision learners," in *CVPR*, 2022.

[163] K. He, H. Fan, Y. Wu, S. Xie, and R. Girshick, "Momentum contrast for unsupervised visual representation learning," in *Proceedings of the IEEE/CVF conference on computer vision and pattern recognition*, pp. 9729–9738, 2020.

[164] K. He, G. Gkioxari, P. Dollár, and R. Girshick, "Mask r-cnn," in *ICCV*, 2017.

[165] K. He, J. Sun, and X. Tang, "Single image haze removal using dark channel prior," *IEEE transactions on pattern analysis and machine intelligence*, vol. 33, no. 12, 2010, pp. 2341–2353.

[166] K. He, X. Zhang, S. Ren, and J. Sun, "Deep residual learning for image recognition," in *CVPR*, 2016.

[167] P. He, X. Liu, J. Gao, and W. Chen, "DeBERTa: Decoding-enhanced bert with disentangled attention," in *ICLR*, 2021.

[168] R. He, S. Sun, X. Yu, C. Xue, W. Zhang, P. Torr, S. Bai, and X. Qi, "Is synthetic data from generative models ready for image recognition?" *arXiv preprint arXiv:2210.07574*, 2022.

[169] O. Henaff, "Data-efficient image recognition with contrastive predictive coding," in *ICML*, 2020.

[170] A. Hertz, R. Mokady, J. Tenenbaum, K. Aberman, Y. Pritch, and D. Cohen-or, "Prompt-to-prompt image editing with cross-attention control," in *The Eleventh International Conference on Learning Representations*, 2022.

[171] R. D. Hjelm, A. Fedorov, S. Lavoie-Marchildon, K. Grewal, P. Bachman, A. Trischler, and Y. Bengio, "Learning deep representations by mutual information estimation and maximization," *arXiv preprint arXiv:1808.06670*, 2018.

[172] J. Ho, W. Chan, C. Saharia, J. Whang, R. Gao, A. Gritsenko, D. P. Kingma, B. Poole, M. Norouzi, D. J. Fleet, *et al.*, "Imagen video: High definition video generation with diffusion models," *arXiv preprint arXiv:2210.02303*, 2022.

[173] J. Ho, A. Jain, and P. Abbeel, "Denoising diffusion probabilistic models," in *NeurIPS*, 2020.

[174] J. Hoffmann, S. Borgeaud, A. Mensch, E. Buchatskaya, T. Cai, E. Rutherford, D. d. L. Casas, L. A. Hendricks, J. Welbl, A. Clark, *et al.*, "Training compute-optimal large language models," *arXiv preprint arXiv:2203.15556*, 2022.

[175] Y. Hong, H. Zhen, P. Chen, S. Zheng, Y. Du, Z. Chen, and C. Gan, "3d-llm: Injecting the 3d world into large language models," *arXiv preprint arXiv:2307.12981*, 2023.

[176] E. J. Hu, Y. Shen, P. Wallis, Z. Allen-Zhu, Y. Li, S. Wang, L. Wang, and W. Chen, "Lora: Low-rank adaptation of large language models," *arXiv preprint arXiv:2106.09685*, 2021.

[177] R. Hu, M. Rohrbach, and T. Darrell, "Segmentation from natural language expressions," in *European Conference on Computer Vision*, Springer, pp. 108–124, 2016.

[178] R. Hu and A. Singh, "Unit: Multimodal multitask learning with a unified transformer," in *Proceedings of the IEEE/CVF International Conference on Computer Vision (ICCV)*, pp. 1439–1449, Oct. 2021.

[179] R. Hu and A. Singh, "Unit: Multimodal multitask learning with a unified transformer," in *ICCV*, 2021.

[180] W. Hu, Y. Xu, Y. Li, W. Li, Z. Chen, and Z. Tu, "Bliva: A simple multimodal llm for better handling of text-rich visual questions," *arXiv preprint arXiv:2308.09936*, 2023.

[181] L. Huang, S. You, M. Zheng, F. Wang, C. Qian, and T. Yamasaki, "Green hierarchical vision transformer for masked image modeling," *NeurIPS*, 2022.

[182] R. Huang, M. Li, D. Yang, J. Shi, X. Chang, Z. Ye, Y. Wu, Z. Hong, J. Huang, J. Liu, *et al.*, "Audiogpt: Understanding and generating speech, music, sound, and talking head," *arXiv preprint arXiv:2304.12995*, 2023.

[183] S. Huang, L. Dong, W. Wang, Y. Hao, S. Singhal, S. Ma, T. Lv, L. Cui, O. K. Mohammed, Q. Liu, *et al.*, "Language is not all you need: Aligning perception with language models," *arXiv preprint arXiv:2302.14045*, 2023.

[184] S. Huang, Z. Jiang, H. Dong, Y. Qiao, P. Gao, and H. Li, "Instruct2act: Mapping multi-modality instructions to robotic actions with large language model," *arXiv preprint arXiv:2305.11176*, 2023.

[185] W. Huang, P. Abbeel, D. Pathak, and I. Mordatch, "Language models as zero-shot planners: Extracting actionable knowledge for embodied agents," in *International Conference on Machine Learning*, PMLR, pp. 9118–9147, 2022.

[186] Y. Huang, Z. Meng, F. Liu, Y. Su, N. Collier, and Y. Lu, "Sparkles: Unlocking chats across multiple images for multimodal instruction-following models," *arXiv preprint arXiv:2308.16463*, 2023.

[187] Z. Huang, Z. Zeng, Y. Huang, B. Liu, D. Fu, and J. Fu, "Seeing out of the box: End-to-end pre-training for vision-language representation learning," in *CVPR*, 2021.

[188] Z. Huang, Z. Zeng, B. Liu, D. Fu, and J. Fu, "Pixel-BERT: Aligning image pixels with text by deep multi-modal transformers," *arXiv preprint arXiv:2004.00849*, 2020.

[189] D. Huynh, J. Kuen, Z. Lin, J. Gu, and E. Elhamifar, "Open-vocabulary instance segmentation via robust cross-modal pseudo-labeling," in *Proceedings of the IEEE/CVF Conference on Computer Vision and Pattern Recognition*, pp. 7020–7031, 2022.

[190] G. Ilharco, M. Wortsman, R. Wightman, C. Gordon, N. Carlini, R. Taori, A. Dave, V. Shankar, H. Namkoong, J. Miller, H. Hajishirzi, A. Farhadi, and L. Schmidt, *Openclip*, version 0.1, Jul. 2021. DOI: 10.5281/zenodo.5143773.

[191] A. Jain, B. Mildenhall, J. T. Barron, P. Abbeel, and B. Poole, "Zero-shot text-guided object generation with dream fields," 2022.

[192] J. Jain, J. Li, M. T. Chiu, A. Hassani, N. Orlov, and H. Shi, "One-former: One transformer to rule universal image segmentation," in *CVPR*, 2023.

[193] A. Jaiswal, A. R. Babu, M. Z. Zadeh, D. Banerjee, and F. Makedon, "A survey on contrastive self-supervised learning," *Technologies*, 2020.

[194] K. R. Jerripothula, J. Cai, and J. Yuan, "Image co-segmentation via saliency co-fusion," *IEEE Transactions on Multimedia*, vol. 18, no. 9, 2016, pp. 1896–1909.

[195] C. Jia, Y. Yang, Y. Xia, Y.-T. Chen, Z. Parekh, H. Pham, Q. V. Le, Y. Sung, Z. Li, and T. Duerig, "Scaling up visual and vision-language representation learning with noisy text supervision," in *ICML*, 2021.

[196] L. Jing and Y. Tian, "Self-supervised visual feature learning with deep neural networks: A survey," *IEEE transactions on pattern analysis and machine intelligence*, 2020.

[197] A. Joulin, F. Bach, and J. Ponce, "Discriminative clustering for image co-segmentation," in *2010 IEEE Computer Society Conference on Computer Vision and Pattern Recognition*, IEEE, pp. 1943–1950, 2010.

[198] A. Kamath, M. Singh, Y. LeCun, G. Synnaeve, I. Misra, and N. Carion, "Mdetr-modulated detection for end-to-end multi-modal understanding," in *ICCV*, 2021.

[199] M. Kang, J.-Y. Zhu, R. Zhang, J. Park, E. Shechtman, S. Paris, and T. Park, "Scaling up gans for text-to-image synthesis," in *Proceedings of the IEEE/CVF Conference on Computer Vision and Pattern Recognition*, pp. 10 124–10 134, 2023.

[200] J. Kaplan, S. McCandlish, T. Henighan, T. B. Brown, B. Chess, R. Child, S. Gray, A. Radford, J. Wu, and D. Amodei, "Scaling laws for neural language models," *arXiv preprint arXiv:2001.08361*, 2020.

[201] B. Kawar, S. Zada, O. Lang, O. Tov, H. Chang, T. Dekel, I. Mosseri, and M. Irani, "Imagic: Text-based real image editing with diffusion models," in *Proceedings of the IEEE/CVF Conference on Computer Vision and Pattern Recognition*, pp. 6007–6017, 2023.

[202] S. Kazemzadeh, V. Ordonez, M. Matten, and T. Berg, "Referitgame: Referring to objects in photographs of natural scenes," in *EMNLP*, 2014.

[203] W. Kim, B. Son, and I. Kim, "ViLT: Vision-and-language transformer without convolution or region supervision," in *ICML*, 2021.

[204] D. P. Kingma and M. Welling, "Auto-encoding variational bayes," *arXiv preprint arXiv:1312.6114*, 2013.

[205] A. Kirillov, K. He, R. Girshick, C. Rother, and P. Dollár, "Panoptic segmentation," in *CVPR*, 2019.

[206] A. Kirillov, E. Mintun, N. Ravi, H. Mao, C. Rolland, L. Gustafson, T. Xiao, S. Whitehead, A. C. Berg, W.-Y. Lo, *et al.*, "Segment anything," *arXiv preprint arXiv:2304.02643*, 2023.

[207] J. Y. Koh, D. Fried, and R. Salakhutdinov, "Generating images with multimodal language models," *arXiv preprint arXiv:2305.17216*, 2023.

[208] T. Kojima, S. S. Gu, M. Reid, Y. Matsuo, and Y. Iwasawa, "Large language models are zero-shot reasoners," *arXiv preprint arXiv:2205.11916*, 2022.

[209] I. Kokkinos, "Ubernet: Training a universal convolutional neural network for low-, mid-, and high-level vision using diverse datasets and limited memory," in *Proceedings of the IEEE conference on computer vision and pattern recognition*, pp. 6129–6138, 2017.

[210] A. Kolesnikov, L. Beyer, X. Zhai, J. Puigcerver, J. Yung, S. Gelly, and N. Houlsby, "Big transfer (bit): General visual representation learning," in *ECCV*, 2020.

[211] A. Kolesnikov, A. S. Pinto, L. Beyer, X. Zhai, J. Harmsen, and N. Houlsby, "Uvim: A unified modeling approach for vision with learned guiding codes," *arXiv preprint arXiv:2205.10337*, 2022.

[212] A. Krizhevsky, I. Sutskever, and G. E. Hinton, "Imagenet classification with deep convolutional neural networks," in *NeurIPS*, 2012.

[213] N. Kumari, B. Zhang, R. Zhang, E. Shechtman, and J.-Y. Zhu, "Multi-concept customization of text-to-image diffusion," in *Proceedings of the IEEE/CVF Conference on Computer Vision and Pattern Recognition*, pp. 1931–1941, 2023.

[214] W. Kuo, F. Bertsch, W. Li, A. Piergiovanni, M. Saffar, and A. Angelova, "Findit: Generalized localization with natural language queries," in *ECCV*, 2022.

[215] X. Lai, Z. Tian, Y. Chen, Y. Li, Y. Yuan, S. Liu, and J. Jia, "Lisa: Reasoning segmentation via large language model," *arXiv preprint arXiv:2308.00692*, 2023.

[216] A. Lamb, V. Dumoulin, and A. Courville, "Discriminative regularization for generative models," *arXiv preprint arXiv:1602.03220*, 2016.

[217] J. Lambert, Z. Liu, O. Sener, J. Hays, and V. Koltun, "Mseg: A composite dataset for multi-domain semantic segmentation," in *Proceedings of the IEEE/CVF conference on computer vision and pattern recognition*, pp. 2879–2888, 2020.

[218] A. B. L. Larsen, S. K. Sønderby, H. Larochelle, and O. Winther, "Autoencoding beyond pixels using a learned similarity metric," in *International conference on machine learning*, PMLR, pp. 1558–1566, 2016.

[219] H. Laurençon, L. Saulnier, L. Tronchon, S. Bekman, A. Singh, A. Lozhkov, T. Wang, S. Karamcheti, A. M. Rush, D. Kiela, *et al.*, "Obelisc: An open web-scale filtered dataset of interleaved image-text documents," *arXiv preprint arXiv:2306.16527*, 2023.

[220] D. Lee, C. Kim, S. Kim, M. Cho, and W.-S. Han, "Autoregressive image generation using residual quantization," in *Proceedings of the IEEE/CVF Conference on Computer Vision and Pattern Recognition*, pp. 11 523–11 532, 2022.

[221] P. Lewis, E. Perez, A. Piktus, F. Petroni, V. Karpukhin, N. Goyal, H. Küttler, M. Lewis, W.-t. Yih, T. Rocktäschel, *et al.*, "Retrieval-augmented generation for knowledge-intensive nlp tasks," in *NeurIPS*, 2020.

[222] B. Li, H. Liu, L. Chen, Y. J. Lee, C. Li, and Z. Liu, "Benchmarking and analyzing generative data for visual recognition," *arXiv preprint arXiv:2307.13697*, 2023.

[223] B. Li, Y. Zhang, L. Chen, J. Wang, F. Pu, J. Yang, C. Li, and Z. Liu, "Mimic-it: Multi-modal in-context instruction tuning," *arXiv preprint arXiv:2306.05425*, 2023.

[224] B. Li, Y. Zhang, L. Chen, J. Wang, J. Yang, and Z. Liu, "Otter: A multi-modal model with in-context instruction tuning," *arXiv preprint arXiv:2305.03726*, 2023.

[225] B. Li, R. Wang, G. Wang, Y. Ge, Y. Ge, and Y. Shan, "Seed-bench: Benchmarking multimodal llms with generative comprehension," *arXiv preprint arXiv:2307.16125*, 2023.

[226] B. Li, K. Q. Weinberger, S. Belongie, V. Koltun, and R. Ranftl, "Language-driven semantic segmentation," in *ICLR*, 2022.

[227] C. Li, H. Liu, L. H. Li, P. Zhang, J. Aneja, J. Yang, P. Jin, Y. J. Lee, H. Hu, Z. Liu, *et al.*, "Elevater: A benchmark and toolkit for evaluating language-augmented visual models," in *NeurIPS, Track on Datasets and Benchmarks*, 2022.

[228] C. Li, C. Wong, S. Zhang, N. Usuyama, H. Liu, J. Yang, T. Naumann, H. Poon, and J. Gao, "Llava-med: Training a large language-and-vision assistant for biomedicine in one day," *arXiv preprint arXiv:2306.00890*, 2023.

[229] C. Li, J. Yang, P. Zhang, M. Gao, B. Xiao, X. Dai, L. Yuan, and J. Gao, "Efficient self-supervised vision transformers for representation learning," *arXiv preprint arXiv:2106.09785*, 2021.

[230] F. Li, H. Zhang, P. Sun, X. Zou, S. Liu, J. Yang, C. Li, L. Zhang, and J. Gao, "Semantic-sam: Segment and recognize anything at any granularity," *arXiv preprint arXiv:2307.04767*, 2023.

[231] F. Li, H. Zhang, Y.-F. Zhang, S. Liu, J. Guo, L. M. Ni, P. Zhang, and L. Zhang, "Vision-language intelligence: Tasks, representation learning, and large models," *arXiv preprint arXiv:2203.01922*, 2022.

[232] G. Li, N. Duan, Y. Fang, M. Gong, and D. Jiang, "Unicoder-vl: A universal encoder for vision and language by cross-modal pre-training," in *AAAI*, 2020.

[233] H. Li, J. Zhu, X. Jiang, X. Zhu, H. Li, C. Yuan, X. Wang, Y. Qiao, X. Wang, W. Wang, *et al.*, "Uni-perceiver v2: A generalist model for large-scale vision and vision-language tasks," in *Proceedings of the IEEE/CVF Conference on Computer Vision and Pattern Recognition*, pp. 2691–2700, 2023.

[234] J. Li, D. Li, S. Savarese, and S. Hoi, "Blip-2: Bootstrapping language-image pre-training with frozen image encoders and large language models," *arXiv preprint arXiv:2301.12597*, 2023.

[235] J. Li, D. Li, C. Xiong, and S. Hoi, "Blip: Bootstrapping language-image pre-training for unified vision-language understanding and generation," in *ICML*, 2022.

[236] J. Li, R. R. Selvaraju, A. D. Gotmare, S. Joty, C. Xiong, and S. Hoi, "Align before fuse: Vision and language representation learning with momentum distillation," in *NeurIPS*, 2021.

[237] K. Li, Y. He, Y. Wang, Y. Li, W. Wang, P. Luo, Y. Wang, L. Wang, and Y. Qiao, "Videochat: Chat-centric video understanding," *arXiv preprint arXiv:2305.06355*, 2023.

[238] L. Li, Y. Yin, S. Li, L. Chen, P. Wang, S. Ren, M. Li, Y. Yang, J. Xu, X. Sun, *et al.*, "M3it: A large-scale dataset towards multi-modal multilingual instruction tuning," *arXiv preprint arXiv:2306.04387*, 2023.

[239] L. Li, Z. Gan, Y. Cheng, and J. Liu, "Relation-aware graph attention network for visual question answering," in *ICCV*, 2019.

[240] L. H. Li, M. Yatskar, D. Yin, C.-J. Hsieh, and K.-W. Chang, "VisualBERT: A simple and performant baseline for vision and language," *arXiv preprint arXiv:1908.03557*, 2019.

[241] L. H. Li, P. Zhang, H. Zhang, J. Yang, C. Li, Y. Zhong, L. Wang, L. Yuan, L. Zhang, J.-N. Hwang, K.-W. Chang, and J. Gao, "Grounded language-image pre-training," *CVPR*, 2022.

[242] L. H. Li, P. Zhang, H. Zhang, J. Yang, C. Li, Y. Zhong, L. Wang, L. Yuan, L. Zhang, J.-N. Hwang, K.-W. Chang, and J. Gao, "Grounded language-image pre-training," in *CVPR*, 2022.

[243] M. Li, F. Song, B. Yu, H. Yu, Z. Li, F. Huang, and Y. Li, "Api-bank: A benchmark for tool-augmented llms," *arXiv preprint arXiv:2304.08244*, 2023.

[244] S. Li and N. Tajbakhsh, "Scigraphqa: A large-scale synthetic multi-turn question-answering dataset for scientific graphs," *arXiv preprint arXiv:2308.03349*, 2023.

[245] X. L. Li and P. Liang, "Prefix-tuning: Optimizing continuous prompts for generation," *arXiv preprint arXiv:2101.00190*, 2021.

[246] X. Li, X. Yin, C. Li, P. Zhang, X. Hu, L. Zhang, L. Wang, H. Hu, L. Dong, F. Wei, Y. Choi, and J. Gao, "Oscar: Object-semantics aligned pre-training for vision-language tasks," in *ECCV*, 2020.

[247] Y. Li, C. Zhang, G. Yu, Z. Wang, B. Fu, G. Lin, C. Shen, L. Chen, and Y. Wei, "Stablellava: Enhanced visual instruction tuning with synthesized image-dialogue data," *arXiv preprint arXiv:2308.10253*, 2023.

[248] Y. Li, F. Liang, L. Zhao, Y. Cui, W. Ouyang, J. Shao, F. Yu, and J. Yan, "Supervision exists everywhere: A data efficient contrastive language-image pre-training paradigm," in *ICLR*, 2022.

[249] Y. Li, H. Fan, R. Hu, C. Feichtenhofer, and K. He, "Scaling language-image pre-training via masking," in *CVPR*, 2023.

[250] Y. Li, Y. Du, K. Zhou, J. Wang, W. X. Zhao, and J.-R. Wen, "Evaluating object hallucination in large vision-language models," *arXiv preprint arXiv:2305.10355*, 2023.

[251] Y. Li, S. Bubeck, R. Eldan, A. D. Giorno, S. Gunasekar, and Y. T. Lee, *Textbooks are all you need ii: Phi-1.5 technical report*, 2023. arXiv: 2309.05463 [cs.CL].

[252] Y. Li, H. Liu, Q. Wu, F. Mu, J. Yang, J. Gao, C. Li, and Y. J. Lee, "Gligen: Open-set grounded text-to-image generation," in *Proceedings of the IEEE/CVF Conference on Computer Vision and Pattern Recognition*, pp. 22 511–22 521, 2023.

[253] F. Liang, B. Wu, X. Dai, K. Li, Y. Zhao, H. Zhang, P. Zhang, P. Vajda, and D. Marculescu, "Open-vocabulary semantic segmentation with mask-adapted clip," in *Proceedings of the IEEE/CVF Conference on Computer Vision and Pattern Recognition*, pp. 7061–7070, 2023.

[254] Y. Liang, C. Wu, T. Song, W. Wu, Y. Xia, Y. Liu, Y. Ou, S. Lu, L. Ji, S. Mao, *et al.*, "Taskmatrix. ai: Completing tasks by connecting foundation models with millions of apis," *arXiv preprint arXiv:2303.16434*, 2023.

[255] B. Lin, B. Zhu, Y. Ye, M. Ning, P. Jin, and L. Yuan, "Video-llava: Learning united visual representation by alignment before projection," *arXiv preprint arXiv:2311.10122*, 2023.

[256] C.-H. Lin, J. Gao, L. Tang, T. Takikawa, X. Zeng, X. Huang, K. Kreis, S. Fidler, M.-Y. Liu, and T.-Y. Lin, "Magic3d: High-resolution text-to-3d content creation," in *IEEE Conference on Computer Vision and Pattern Recognition (CVPR)*, 2023.

[257] T.-Y. Lin, P. Goyal, R. Girshick, K. He, and P. Dollár, "Focal loss for dense object detection," in *Proceedings of the IEEE international conference on computer vision*, pp. 2980–2988, 2017.

[258] T.-Y. Lin, M. Maire, S. Belongie, J. Hays, P. Perona, D. Ramanan, P. Dollár, and C. L. Zitnick, "Microsoft coco: Common objects in context," in *ECCV*, 2014.

[259] Y. Lin, H. Wu, R. Wang, H. Lu, X. Lin, H. Xiong, and L. Wang, "Towards language-guided interactive 3d generation: Llms as layout interpreter with generative feedback," *arXiv preprint arXiv:2305.15808*, 2023.

[260] C. Liu, Z. Lin, X. Shen, J. Yang, X. Lu, and A. Yuille, "Recurrent multimodal interaction for referring image segmentation," in *Proceedings of the IEEE International Conference on Computer Vision*, pp. 1271–1280, 2017.

[261] F. Liu, K. Lin, L. Li, J. Wang, Y. Yacoob, and L. Wang, "Aligning large multi-modal model with robust instruction tuning," *arXiv preprint arXiv:2306.14565*, 2023.

[262] H. Liu, X. Jiang, X. Li, A. Guo, Y. Hu, D. Jiang, and B. Ren, "The devil is in the frequency: Geminated gestalt autoencoder for self-supervised visual pre-training," in *AAAI*, 2023.

[263] H. Liu, C. Li, Y. Li, and Y. J. Lee, *Improved baselines with visual instruction tuning*, 2023.

[264] H. Liu, C. Li, Q. Wu, and Y. J. Lee, "Visual instruction tuning," *arXiv preprint arXiv:2304.08485*, 2023.

[265] H. Liu, K. Son, J. Yang, C. Liu, J. Gao, Y. J. Lee, and C. Li, "Learning customized visual models with retrieval-augmented knowledge," in *Proceedings of the IEEE/CVF Conference on Computer Vision and Pattern Recognition*, 2023.

[266] J. Liu, H. Ding, Z. Cai, Y. Zhang, R. K. Satzoda, V. Mahadevan, and R. Manmatha, *Polyformer: Referring image segmentation as sequential polygon generation*, 2023. arXiv: 2302.07387 [cs.CV].

[267] N. Liu, S. Li, Y. Du, A. Torralba, and J. B. Tenenbaum, "Compositional visual generation with composable diffusion models," in *European Conference on Computer Vision*, Springer, pp. 423–439, 2022.

[268] R. Liu, R. Wu, B. V. Hoorick, P. Tokmakov, S. Zakharov, and C. Vondrick, *Zero-1-to-3: Zero-shot one image to 3d object*, 2023. arXiv: 2303.11328 [cs.CV].

[269] S. Liu, L. Fan, E. Johns, Z. Yu, C. Xiao, and A. Anandkumar, *Prismer: A vision-language model with an ensemble of experts*, 2023. arXiv: 2303.02506 [cs.LG].

[270] S. Liu, H. Cheng, H. Liu, H. Zhang, F. Li, T. Ren, X. Zou, J. Yang, H. Su, J. Zhu, L. Zhang, J. Gao, and C. Li, *Llava-plus: Learning to use tools for creating multimodal agents*, 2023.

[271] S. Liu, Z. Zeng, T. Ren, F. Li, H. Zhang, J. Yang, C. Li, J. Yang, H. Su, J. Zhu, *et al.*, "Grounding dino: Marrying dino with grounded pre-training for open-set object detection," *arXiv preprint arXiv:2303.05499*, 2023.

[272] S. Liu, J. Ye, and X. Wang, "Any-to-any style transfer," *arXiv preprint arXiv:2304.09728*, 2023.

[273] W. Liu, D. Anguelov, D. Erhan, C. Szegedy, S. Reed, C.-Y. Fu, and A. C. Berg, "Ssd: Single shot multibox detector," in *Computer Vision–ECCV 2016: 14th European Conference, Amsterdam, The Netherlands, October 11–14, 2016, Proceedings, Part I 14*, Springer, pp. 21–37, 2016.

[274] W. Liu and Y. Zuo, "Stone needle: A general multimodal large-scale model framework towards healthcare," *arXiv preprint arXiv:2306.16034*, 2023.

[275] X. Liu, C. Gong, and Q. Liu, *Flow straight and fast: Learning to generate and transfer data with rectified flow*, 2022. arXiv: 2209.03003 [cs.LG].

[276] X. Liu, Z. Zhu, H. Liu, Y. Yuan, M. Cui, Q. Huang, J. Liang, Y. Cao, Q. Kong, M. D. Plumbley, *et al.*, "Wavjourney: Compositional audio creation with large language models," *arXiv preprint arXiv:2307.14335*, 2023.

[277] Y. Liu, M. Ott, N. Goyal, J. Du, M. Joshi, D. Chen, O. Levy, M. Lewis, L. Zettlemoyer, and V. Stoyanov, "Roberta: A robustly optimized bert pretraining approach," *arXiv preprint arXiv:1907.11692*, 2019.

[278] Y. Liu, H. Duan, Y. Zhang, B. Li, S. Zhang, W. Zhao, Y. Yuan, J. Wang, C. He, Z. Liu, *et al.*, "Mmbench: Is your multi-modal model an all-around player?" *arXiv preprint arXiv:2307.06281*, 2023.

[279] Y. Liu, Z. Li, H. Li, W. Yu, M. Huang, D. Peng, M. Liu, M. Chen, C. Li, L. Jin, *et al.*, "On the hidden mystery of ocr in large multimodal models," *arXiv preprint arXiv:2305.07895*, 2023.

[280] Z. Liu, Y. Lin, Y. Cao, H. Hu, Y. Wei, Z. Zhang, S. Lin, and B. Guo, "Swin transformer: Hierarchical vision transformer using shifted windows," in *ICCV*, 2021.

[281] Z. Liu, Y. He, W. Wang, W. Wang, Y. Wang, S. Chen, Q. Zhang, Y. Yang, Q. Li, J. Yu, *et al.*, "Internchat: Solving vision-centric tasks by interacting with chatbots beyond language," *arXiv preprint arXiv:2305.05662*, 2023.

[282] Z. Liu, H. Mao, C.-Y. Wu, C. Feichtenhofer, T. Darrell, and S. Xie, "A convnet for the 2020s," in *Proceedings of the IEEE/CVF conference on computer vision and pattern recognition*, pp. 11 976–11 986, 2022.

[283] A. Long, W. Yin, T. Ajanthan, V. Nguyen, P. Purkait, R. Garg, A. Blair, C. Shen, and A. van den Hengel, "Retrieval augmented classification for long-tail visual recognition," in *CVPR*, 2022.

[284] J. Long, E. Shelhamer, and T. Darrell, "Fully convolutional networks for semantic segmentation," in *CVPR*, 2015.

[285] C.-Z. Lu, X. Jin, Q. Hou, J. H. Liew, M.-M. Cheng, and J. Feng, "Delving deeper into data scaling in masked image modeling," *arXiv preprint arXiv:2305.15248*, 2023.

[286] J. Lu, D. Batra, D. Parikh, and S. Lee, "Vilbert: Pretraining task-agnostic visiolinguistic representations for vision-and-language tasks," in *NeurIPS*, 2019.

[287] J. Lu, C. Clark, R. Zellers, R. Mottaghi, and A. Kembhavi, "Unified-io: A unified model for vision, language, and multi-modal tasks," *arXiv preprint arXiv:2206.08916*, 2022.

[288] J. Lu, V. Goswami, M. Rohrbach, D. Parikh, and S. Lee, "12-in-1: Multi-task vision and language representation learning," in *Proceedings of the IEEE/CVF conference on computer vision and pattern recognition*, pp. 10 437–10 446, 2020.

[289] P. Lu, H. Bansal, T. Xia, J. Liu, C. Li, H. Hajishirzi, H. Cheng, K.-W. Chang, M. Galley, and J. Gao, "Mathvista: Evaluating mathematical reason-ing of foundation models in visual contexts," *arXiv preprint arXiv:2310.02255*, 2023.

[290] P. Lu, S. Mishra, T. Xia, L. Qiu, K.-W. Chang, S.-C. Zhu, O. Tafjord, P. Clark, and A. Kalyan, "Learn to explain: Multimodal reasoning via thought chains for science question answering," *Advances in Neural Information Processing Systems*, 2022.

[291] P. Lu, B. Peng, H. Cheng, M. Galley, K.-W. Chang, Y. N. Wu, S.-C. Zhu, and J. Gao, "Chameleon: Plug-and-play compositional reasoning with large language models," *arXiv preprint arXiv:2304.09842*, 2023.

[292] Q. Lu, J. Kuen, S. Tiancheng, G. Jiuxiang, G. Weidong, J. Jiaya, L. Zhe, and Y. Ming-Hsuan, "High-quality entity segmentation," in *ICCV*, 2023.

[293] Y. Lu, C. Li, H. Liu, J. Yang, J. Gao, and Y. Shen, "An empirical study of scaling instruction-tuned large multimodal models," *arXiv preprint*, 2023.

[294] T. Lüddecke and A. Ecker, "Image segmentation using text and image prompts," in *CVPR*, 2022.

[295] G. Luo, Y. Zhou, T. Ren, S. Chen, X. Sun, and R. Ji, "Cheap and quick: Efficient vision-language instruction tuning for large language models," *arXiv preprint arXiv:2305.15023*, 2023.

[296] H. Luo, J. Bao, Y. Wu, X. He, and T. Li, "Segclip: Patch aggregation with learnable centers for open-vocabulary semantic segmentation," in *International Conference on Machine Learning*, PMLR, pp. 23 033–23 044, 2023.

[297] R. Luo, Z. Zhao, M. Yang, J. Dong, M. Qiu, P. Lu, T. Wang, and Z. Wei, "Valley: Video assistant with large language model enhanced ability," *arXiv preprint arXiv:2306.07207*, 2023.

[298] W. Luo, J. Xing, A. Milan, X. Zhang, W. Liu, and T.-K. Kim, "Multiple object tracking: A literature review," *Artificial intelligence*, vol. 293, 2021, p. 103 448.

[299] J. Ma and B. Wang, "Segment anything in medical images," *arXiv preprint arXiv:2304.12306*, 2023.

[300] Z. Ma, X. Hong, and Q. Shangguan, "Can sam count anything? an empirical study on sam counting," *arXiv preprint arXiv:2304.10817*, 2023.

[301] J. Mao, J. Huang, A. Toshev, O. Camburu, A. L. Yuille, and K. Murphy, "Generation and comprehension of unambiguous object descriptions," in *CVPR*, 2016.

[302] E. Margffoy-Tuay, J. C. Pérez, E. Botero, and P. Arbeláez, "Dynamic multimodal instance segmentation guided by natural language queries," in *Proceedings of the European Conference on Computer Vision (ECCV)*, pp. 630–645, 2018.

[303] K. Marino, X. Chen, D. Parikh, A. Gupta, and M. Rohrbach, "Krisp: Integrating implicit and symbolic knowledge for open-domain knowledge-based vqa," in *CVPR*, 2021.

[304] M. Mazumder, C. Banbury, X. Yao, B. Karlaš, W. G. Rojas, S. Diamos, G. Diamos, L. He, D. Kiela, D. Jurado, *et al.*, "Dataperf: Benchmarks for data-centric ai development," *arXiv preprint arXiv:2207.10062*, 2022.

[305] K. McGuinness and N. E. O'connor, "A comparative evaluation of interactive segmentation algorithms," *Pattern Recognition*, vol. 43, no. 2, 2010, pp. 434–444.

[306] C. Meng, Y. He, Y. Song, J. Song, J. Wu, J.-Y. Zhu, and S. Ermon, "Sdedit: Guided image synthesis and editing with stochastic differential equations," in *International Conference on Learning Representations*, 2021.

[307] A. Mertan, D. J. Duff, and G. Unal, "Single image depth estimation: An overview," *Digital Signal Processing*, vol. 123, 2022, p. 103 441.

[308] A. Miech, D. Zhukov, J.-B. Alayrac, M. Tapaswi, I. Laptev, and J. Sivic, "Howto100m: Learning a text-video embedding by watching hundred million narrated video clips," in *ICCV*, 2019.

[309] T. Mikolov, K. Chen, G. Corrado, and J. Dean, "Efficient estimation of word representations in vector space," *arXiv preprint arXiv:1301.3781*, 2013.

[310] M. Minderer, A. Gritsenko, A. Stone, M. Neumann, D. Weissenborn, A. Dosovitskiy, A. Mahendran, A. Arnab, M. Dehghani, Z. Shen, X. Wang, X. Zhai, T. Kipf, and N. Houlsby, *Simple open-vocabulary object detection with vision transformers*, 2022. arXiv: 2205.06230 [cs.CV].

[311] I. Misra and L. v. d. Maaten, "Self-supervised learning of pretext-invariant representations," in *CVPR*, 2020.

[312] I. Misra, A. Shrivastava, A. Gupta, and M. Hebert, "Cross-stitch networks for multi-task learning," in *Proceedings of the IEEE conference on computer vision and pattern recognition*, pp. 3994–4003, 2016.

[313] M. Monajatipoor, L. H. Li, M. Rouhsedaghat, L. F. Yang, and K.-W. Chang, "Metavl: Transferring in-context learning ability from language models to vision-language models," *arXiv preprint arXiv:2306.01311*, 2023.

[314] *Moonvalley*, 2023. URL: https://moonvalley.ai/.

[315] M. Moor, Q. Huang, S. Wu, M. Yasunaga, C. Zakka, Y. Dalmia, E. P. Reis, P. Rajpurkar, and J. Leskovec, "Med-flamingo: A multimodal medical few-shot learner," *arXiv preprint arXiv:2307.15189*, 2023.

[316] *Morph*, 2023. URL: https://www.morphstudio.com/.

[317] E. N. Mortensen and W. A. Barrett, "Interactive segmentation with intelligent scissors," *Graphical models and image processing*, vol. 60, no. 5, 1998, pp. 349–384.

[318] MosaicML NLP Team, *Introducing mpt-7b: A new standard for open-source, ly usable llms*, 2023. URL: www.mosaicml.com/blog/mpt-7b.

[319] R. Mottaghi, X. Chen, X. Liu, N.-G. Cho, S.-W. Lee, S. Fidler, R. Urtasun, and A. Yuille, "The role of context for object detection and semantic segmentation in the wild," in *Proceedings of the IEEE conference on computer vision and pattern recognition*, pp. 891–898, 2014.

[320] C. Mou, X. Wang, L. Xie, J. Zhang, Z. Qi, Y. Shan, and X. Qie, "T2i-adapter: Learning adapters to dig out more controllable ability for text-to-image diffusion models," *arXiv preprint arXiv:2302.08453*, 2023.

[321] N. Mu, A. Kirillov, D. Wagner, and S. Xie, "Slip: Self-supervision meets language-image pre-training," *arXiv preprint arXiv:2112.12750*, 2021.

[322] Y. Mu, Q. Zhang, M. Hu, W. Wang, M. Ding, J. Jin, B. Wang, J. Dai, Y. Qiao, and P. Luo, "Embodiedgpt: Vision-language pre-training via embodied chain of thought," *arXiv preprint arXiv:2305.15021*, 2023.

[323] S. Munasinghe, R. Thushara, M. Maaz, H. A. Rasheed, S. Khan, M. Shah, and F. Khan, *Pg-video-llava: Pixel grounding large video-language models*, 2023. arXiv: 2311.13435 [cs.CV].

[324] K. Musgrave, S. Belongie, and S.-N. Lim, "A metric learning reality check," in *ECCV*, 2020.

[325] V. K. Nagaraja, V. I. Morariu, and L. S. Davis, "Modeling context between objects for referring expression understanding," in *ECCV*, 2016.

[326] R. Nakano, J. Hilton, S. Balaji, J. Wu, L. Ouyang, C. Kim, C. Hesse, S. Jain, V. Kosaraju, W. Saunders, *et al.*, "Webgpt: Browser-assisted question-answering with human feedback," *arXiv preprint arXiv:2112.09332*, 2021.

[327] T. Nguyen, G. Ilharco, M. Wortsman, S. Oh, and L. Schmidt, "Quality not quantity: On the interaction between dataset design and robustness of clip," *NeurIPS*, 2022.

[328] J. Ning, C. Li, Z. Zhang, Z. Geng, Q. Dai, K. He, and H. Hu, "All in tokens: Unifying output space of visual tasks via soft token," *arXiv preprint arXiv:2301.02229*, 2023.

[329] A. El-Nouby, G. Izacard, H. Touvron, I. Laptev, H. Jegou, and E. Grave, "Are large-scale datasets necessary for self-supervised pre-training?" *arXiv preprint arXiv:2112.10740*, 2021.
[330] A. van den Oord, O. Vinyals, and K. Kavukcuoglu, "Neural discrete representation learning," in *NeurIPS*, 2017.
[331] A. v. d. Oord, Y. Li, and O. Vinyals, "Representation learning with contrastive predictive coding," *arXiv preprint arXiv:1807.03748*, 2018.
[332] A. v. d. Oord, O. Vinyals, and K. Kavukcuoglu, "Neural discrete representation learning," *arXiv preprint arXiv:1711.00937*, 2017.
[333] OpenAI, *ChatGPT*, 2022. URL: https://openai.com/blog/chatgpt/.
[334] OpenAI, *GPT-4 technical report*, 2023. URL: https://arxiv.org/abs/2303.08774.
[335] OpenAI, *Gpt-4 technical report*, 2023. arXiv: 2303.08774 [cs.CL].
[336] M. Oquab, T. Darcet, T. Moutakanni, H. Vo, M. Szafraniec, V. Khalidov, P. Fernandez, D. Haziza, F. Massa, A. El-Nouby, *et al.*, "Dinov2: Learning robust visual features without supervision," *arXiv preprint arXiv:2304.07193*, 2023.
[337] V. Ordonez, G. Kulkarni, and T. Berg, "Im2text: Describing images using 1 million captioned photographs," in *NeurIPS*, 2011.
[338] L. Ouyang, J. Wu, X. Jiang, D. Almeida, C. Wainwright, P. Mishkin, C. Zhang, S. Agarwal, K. Slama, A. Ray, *et al.*, "Training language models to follow instructions with human feedback," *Advances in Neural Information Processing Systems*, vol. 35, 2022, pp. 27 730–27 744.
[339] U. Ozbulak, H. J. Lee, B. Boga, E. T. Anzaku, H. Park, A. Van Messem, W. De Neve, and J. Vankerschaver, "Know your self-supervised learning: A survey on image-based generative and discriminative training," *arXiv preprint arXiv:2305.13689*, 2023.
[340] B. Paranjape, S. Lundberg, S. Singh, H. Hajishirzi, L. Zettlemoyer, and M. T. Ribeiro, "Art: Automatic multi-step reasoning and tool-use for large language models," *arXiv preprint arXiv:2303.09014*, 2023.

[341] S. G. Patil, T. Zhang, X. Wang, and J. E. Gonzalez, "Gorilla: Large language model connected with massive apis," *arXiv preprint arXiv:2305.15334*, 2023.

[342] W. Peebles and S. Xie, "Scalable diffusion models with transformers," *arXiv preprint arXiv:2212.09748*, 2022.

[343] G. Penedo, Q. Malartic, D. Hesslow, R. Cojocaru, A. Cappelli, H. Alobeidli, B. Pannier, E. Almazrouei, and J. Launay, "The RefinedWeb dataset for Falcon LLM: Outperforming curated corpora with web data, and web data only," *arXiv preprint arXiv:2306.01116*, 2023. URL: https://arxiv.org/abs/2306.01116.

[344] B. Peng, M. Galley, P. He, H. Cheng, Y. Xie, Y. Hu, Q. Huang, L. Liden, Z. Yu, W. Chen, and J. Gao, "Check your facts and try again: Improving large language models with external knowledge and automated feedback," *arXiv preprint arXiv:2302.12813*, 2023.

[345] B. Peng, C. Li, P. He, M. Galley, and J. Gao, "Instruction tuning with GPT-4," *arXiv preprint arXiv:2304.03277*, 2023.

[346] Z. Peng, L. Dong, H. Bao, Q. Ye, and F. Wei, "A unified view of masked image modeling," *arXiv preprint arXiv:2210.10615*, 2022.

[347] Z. Peng, L. Dong, H. Bao, Q. Ye, and F. Wei, "Beit v2: Masked image modeling with vector-quantized visual tokenizers," *arXiv preprint arXiv:2208.06366*, 2022.

[348] Z. Peng, W. Wang, L. Dong, Y. Hao, S. Huang, S. Ma, and F. Wei, "Kosmos-2: Grounding multimodal large language models to the world," *arXiv preprint arXiv:2306.14824*, 2023.

[349] M. E. Peters, M. Neumann, R. L. Logan IV, R. Schwartz, V. Joshi, S. Singh, and N. A. Smith, "Knowledge enhanced contextual word representations," *arXiv preprint arXiv:1909.04164*, 2019.

[350] H. Pham, Z. Dai, G. Ghiasi, H. Liu, A. W. Yu, M.-T. Luong, M. Tan, and Q. V. Le, "Combined scaling for zero-shot transfer learning," *arXiv preprint arXiv:2111.10050*, 2021.

[351] R. Pi, J. Gao, S. Diao, R. Pan, H. Dong, J. Zhang, L. Yao, J. Han, H. Xu, and L. K. T. Zhang, "Detgpt: Detect what you need via reasoning," *arXiv preprint arXiv:2305.14167*, 2023.

[352] *Pika 1.0*, 2023. URL: https://pika.art/.

[353] B. A. Plummer, L. Wang, C. M. Cervantes, J. C. Caicedo, J. Hockenmaier, and S. Lazebnik, "Flickr30k entities: Collecting region-to-phrase correspondences for richer image-to-sentence models," in *ICCV*, 2015.

[354] J. Pont-Tuset, J. Uijlings, S. Changpinyo, R. Soricut, and V. Ferrari, "Connecting vision and language with localized narratives," in *ECCV*, 2020.

[355] B. Poole, A. Jain, J. T. Barron, and B. Mildenhall, "Dreamfusion: Text-to-3d using 2d diffusion," *arXiv*, 2022.

[356] C. Qian, C. Han, Y. R. Fung, Y. Qin, Z. Liu, and H. Ji, "Creator: Disentangling abstract and concrete reasonings of large language models through tool creation," *arXiv preprint arXiv:2305.14318*, 2023.

[357] R. Qian, Y. Li, Z. Xu, M.-H. Yang, S. Belongie, and Y. Cui, "Multimodal open-vocabulary video classification via pre-trained vision and language models," *arXiv preprint arXiv:2207.07646*, 2022.

[358] C. Qin, S. Zhang, N. Yu, Y. Feng, X. Yang, Y. Zhou, H. Wang, J. C. Niebles, C. Xiong, S. Savarese, *et al.*, "Unicontrol: A unified diffusion model for controllable visual generation in the wild," *arXiv preprint arXiv:2305.11147*, 2023.

[359] J. Qin, J. Wu, P. Yan, M. Li, R. Yuxi, X. Xiao, Y. Wang, R. Wang, S. Wen, X. Pan, and X. Wang, *Freeseg: Unified, universal and open-vocabulary image segmentation*, 2023. arXiv: 2303.17225 `[cs.CV]`.

[360] A. Radford, J. W. Kim, C. Hallacy, A. Ramesh, G. Goh, S. Agarwal, G. Sastry, A. Askell, P. Mishkin, J. Clark, *et al.*, "Learning transferable visual models from natural language supervision," in *ICML*, 2021.

[361] A. Radford, J. W. Kim, T. Xu, G. Brockman, C. McLeavey, and I. Sutskever, *Robust speech recognition via large-scale weak supervision*, 2022. arXiv: 2212.04356 `[eess.AS]`.

[362] A. Radford, J. Wu, R. Child, D. Luan, D. Amodei, I. Sutskever, *et al.*, "Language models are unsupervised multitask learners," *OpenAI blog*, 2019.

[363] C. Raffel, N. Shazeer, A. Roberts, K. Lee, S. Narang, M. Matena, Y. Zhou, W. Li, and P. J. Liu, "Exploring the limits of transfer learning with a unified text-to-text transformer," *JMLR*, 2020.

[364] S. Rahman, S. Khan, and N. Barnes, "Improved visual-semantic alignment for zero-shot object detection," in *Proceedings of the AAAI Conference on Artificial Intelligence*, vol. 34, pp. 11 932–11 939, 2020.

[365] F. Rajič, L. Ke, Y.-W. Tai, C.-K. Tang, M. Danelljan, and F. Yu, "Segment anything meets point tracking," *arXiv preprint arXiv:2307.01197*, 2023.

[366] A. Ramesh, P. Dhariwal, A. Nichol, C. Chu, and M. Chen, "Hierarchical text-conditional image generation with clip latents," *arXiv preprint arXiv:2204.06125*, 2022.

[367] A. Ramesh, M. Pavlov, G. Goh, S. Gray, C. Voss, A. Radford, M. Chen, and I. Sutskever, "Zero-Shot Text-to-Image Generation," in *ICML*, 2021.

[368] A. Ramesh, M. Pavlov, G. Goh, S. Gray, C. Voss, A. Radford, M. Chen, and I. Sutskever, "Zero-shot text-to-image generation," in *International Conference on Machine Learning*, PMLR, pp. 8821–8831, 2021.

[369] Y. Rao, W. Zhao, G. Chen, Y. Tang, Z. Zhu, G. Huang, J. Zhou, and J. Lu, "Denseclip: Language-guided dense prediction with context-aware prompting," in *Proceedings of the IEEE/CVF Conference on Computer Vision and Pattern Recognition*, pp. 18 082–18 091, 2022.

[370] A. Razavi, A. Van den Oord, and O. Vinyals, "Generating diverse high-fidelity images with vq-vae-2," in *NeurIPS*, 2019.

[371] J. Redmon, S. Divvala, R. Girshick, and A. Farhadi, "You only look once: Unified, real-time object detection," in *Proceedings of the IEEE conference on computer vision and pattern recognition*, pp. 779–788, 2016.

[372] S. Reed, K. Zolna, E. Parisotto, S. G. Colmenarejo, A. Novikov, G. Barth-Maron, M. Gimenez, Y. Sulsky, J. Kay, J. T. Springenberg, *et al.*, "A generalist agent," *arXiv preprint arXiv:2205.06175*, 2022.

[373] S. Ren, K. He, R. Girshick, and J. Sun, "Faster r-cnn: Towards real-time object detection with region proposal networks," in *NeurIPS*, 2015.

[374] T. Ridnik, E. Ben-Baruch, A. Noy, and L. Zelnik-Manor, "Imagenet-21k pretraining for the masses," *arXiv preprint arXiv:2104.10972*, 2021.

[375] R. Rombach, A. Blattmann, D. Lorenz, P. Esser, and B. Ommer, *High-resolution image synthesis with latent diffusion models*, 2021. arXiv: 2112.10752 [cs.CV].

[376] R. Rombach, A. Blattmann, D. Lorenz, P. Esser, and B. Ommer, "High-resolution image synthesis with latent diffusion models," in *CVPR*, 2022.

[377] O. Ronneberger, P. Fischer, and T. Brox, "U-net: Convolutional networks for biomedical image segmentation," in *Medical Image Computing and Computer-Assisted Intervention–MICCAI 2015: 18th International Conference, Munich, Germany, October 5-9, 2015, Proceedings, Part III 18*, Springer, pp. 234–241, 2015.

[378] S. Roy, T. Wald, G. Koehler, M. R. Rokuss, N. Disch, J. Holzschuh, D. Zimmerer, and K. H. Maier-Hein, "Sam. md: Zero-shot medical image segmentation capabilities of the segment anything model," *arXiv preprint arXiv:2304.05396*, 2023.

[379] L. Ruan and Q. Jin, "Survey: Transformer based video-language pre-training," *AI Open*, 2022.

[380] N. Ruiz, Y. Li, V. Jampani, Y. Pritch, M. Rubinstein, and K. Aberman, "Dreambooth: Fine tuning text-to-image diffusion models for subject-driven generation," in *Proceedings of the IEEE/CVF Conference on Computer Vision and Pattern Recognition*, pp. 22 500–22 510, 2023.

[381] O. Russakovsky, J. Deng, H. Su, J. Krause, S. Satheesh, S. Ma, Z. Huang, A. Karpathy, A. Khosla, M. Bernstein, *et al.*, "Imagenet large scale visual recognition challenge," *IJCV*, 2015.

[382] C. Saharia, W. Chan, S. Saxena, L. Li, J. Whang, E. Denton, S. K. S. Ghasemipour, B. K. Ayan, S. S. Mahdavi, R. G. Lopes, *et al.*, "Photorealistic text-to-image diffusion models with deep language understanding," *arXiv preprint arXiv:2205.11487*, 2022.

[383] M. B. Sariyildiz, J. Perez, and D. Larlus, “Learning visual representations with caption annotations,” in *ECCV*, 2020.
[384] T. Schick, J. Dwivedi-Yu, R. Dessì, R. Raileanu, M. Lomeli, L. Zettlemoyer, N. Cancedda, and T. Scialom, “Toolformer: Language models can teach themselves to use tools,” *arXiv preprint arXiv:2302.04761*, 2023.
[385] C. Schuhmann, R. Beaumont, R. Vencu, C. Gordon, R. Wightman, M. Cherti, T. Coombes, A. Katta, C. Mullis, M. Wortsman, *et al.*, “Laion-5b: An open large-scale dataset for training next generation image-text models,” *NeurIPS*, 2022.
[386] C. Schuhmann, R. Vencu, R. Beaumont, R. Kaczmarczyk, C. Mullis, A. Katta, T. Coombes, J. Jitsev, and A. Komatsuzaki, “Laion-400m: Open dataset of clip-filtered 400 million image-text pairs,” *arXiv preprint arXiv:2111.02114*, 2021.
[387] D. Schwenk, A. Khandelwal, C. Clark, K. Marino, and R. Mottaghi, “A-okvqa: A benchmark for visual question answering using world knowledge,” *arXiv preprint arXiv:2206.01718*, 2022.
[388] R. Sennrich, B. Haddow, and A. Birch, “Neural machine translation of rare words with subword units,” in *ACL*, 2016.
[389] S. Shao, Z. Li, T. Zhang, C. Peng, G. Yu, X. Zhang, J. Li, and J. Sun, “Objects365: A large-scale, high-quality dataset for object detection,” in *ICCV*, 2019.
[390] W. Shao, Y. Hu, P. Gao, M. Lei, K. Zhang, F. Meng, P. Xu, S. Huang, H. Li, Y. Qiao, *et al.*, “Tiny lvlm-ehub: Early multimodal experiments with bard,” *arXiv preprint arXiv:2308.03729*, 2023.
[391] ShareGPT, 2023. URL: https://sharegpt.com/.
[392] P. Sharma, N. Ding, S. Goodman, and R. Soricut, “Conceptual captions: A cleaned, hypernymed, image alt-text dataset for automatic image captioning,” in *ACL*, 2018.
[393] Q. Shen, X. Yang, and X. Wang, “Anything-3d: Towards single-view anything reconstruction in the wild,” *arXiv preprint arXiv:2304.10261*, 2023.
[394] S. Shen, C. Li, X. Hu, Y. Xie, J. Yang, P. Zhang, A. Rohrbach, Z. Gan, L. Wang, L. Yuan, C. Liu, K. Keutzer, T. Darrell, A. Rohrbach, and J. Gao, “K-lite: Learning transferable visual models with external knowledge,” in *NeurIPS*, 2022.

[395] S. Shen, L. H. Li, H. Tan, M. Bansal, A. Rohrbach, K.-W. Chang, Z. Yao, and K. Keutzer, "How much can clip benefit vision-and-language tasks?" In *ICLR*, 2022.

[396] Y. Shen, K. Song, X. Tan, D. Li, W. Lu, and Y. Zhuang, "Hugginggpt: Solving ai tasks with chatgpt and its friends in huggingface," *arXiv preprint arXiv:2303.17580*, 2023.

[397] S. Sheynin, O. Ashual, A. Polyak, U. Singer, O. Gafni, E. Nachmani, and Y. Taigman, "Knn-diffusion: Image generation via large-scale retrieval," *arXiv preprint arXiv:2204.02849*, 2022.

[398] J. Shi, W. Xiong, Z. Lin, and H. J. Jung, "Instantbooth: Personalized text-to-image generation without test-time finetuning," *arXiv preprint arXiv:2304.03411*, 2023.

[399] P. Shi, J. Qiu, S. M. D. Abaxi, H. Wei, F. P.-W. Lo, and W. Yuan, "Generalist vision foundation models for medical imaging: A case study of segment anything model on zero-shot medical segmentation," *Diagnostics*, 2023.

[400] T. Shin, Y. Razeghi, R. L. Logan IV, E. Wallace, and S. Singh, "Autoprompt: Eliciting knowledge from language models with automatically generated prompts," *arXiv preprint arXiv:2010.15980*, 2020.

[401] O. Sidorov, R. Hu, M. Rohrbach, and A. Singh, "Textcaps: A dataset for image captioning with reading comprehension," in *ECCV*, 2020.

[402] N. Silberman, D. Hoiem, P. Kohli, and R. Fergus, "Indoor segmentation and support inference from rgbd images," in *Computer Vision–ECCV 2012: 12th European Conference on Computer Vision, Florence, Italy, October 7-13, 2012, Proceedings, Part V 12*, Springer, pp. 746–760, 2012.

[403] U. Singer, A. Polyak, T. Hayes, X. Yin, J. An, S. Zhang, Q. Hu, H. Yang, O. Ashual, O. Gafni, *et al.*, "Make-a-video: Text-to-video generation without text-video data," *arXiv preprint arXiv:2209.14792*, 2022.

[404] A. Singh, R. Hu, V. Goswami, G. Couairon, W. Galuba, M. Rohrbach, and D. Kiela, "Flava: A foundational language and vision alignment model," in *CVPR*, 2022.

[405] M. Singh, Q. Duval, K. V. Alwala, H. Fan, V. Aggarwal, A. Adcock, A. Joulin, P. Dollár, C. Feichtenhofer, R. Girshick, *et al.*, "The effectiveness of mae pre-pretraining for billion-scale pretraining," *arXiv preprint arXiv:2303.13496*, 2023.

[406] M. Singh, L. Gustafson, A. Adcock, V. de Freitas Reis, B. Gedik, R. P. Kosaraju, D. Mahajan, R. Girshick, P. Dollár, and L. Van Der Maaten, "Revisiting weakly supervised pre-training of visual perception models," in *CVPR*, 2022.

[407] J. Sohl-Dickstein, E. Weiss, N. Maheswaranathan, and S. Ganguli, "Deep unsupervised learning using nonequilibrium thermodynamics," in *International conference on machine learning*, PMLR, pp. 2256–2265, 2015.

[408] Y. Song, P. Dhariwal, M. Chen, and I. Sutskever, *Consistency models*, 2023. arXiv: 2303.01469 [cs.LG].

[409] Y. Song and S. Ermon, "Improved techniques for training score-based generative models," *Advances in neural information processing systems*, vol. 33, 2020, pp. 12 438–12 448.

[410] Y. Song, W. Xiong, D. Zhu, C. Li, K. Wang, Y. Tian, and S. Li, "Restgpt: Connecting large language models with real-world applications via restful apis," *arXiv preprint arXiv:2306.06624*, 2023.

[411] K. Srinivasan, K. Raman, J. Chen, M. Bendersky, and M. Najork, "Wit: Wikipedia-based image text dataset for multimodal multilingual machine learning," *arXiv preprint arXiv:2103.01913*, 2021.

[412] *Stable diffusion*, 2022. URL: https://github.com/CompVis/stable-diffusion.

[413] W. Su, X. Zhu, Y. Cao, B. Li, L. Lu, F. Wei, and J. Dai, "VL-BERT: Pre-training of generic visual-linguistic representations," in *ICLR*, 2019.

[414] Y. Su, T. Lan, H. Li, J. Xu, Y. Wang, and D. Cai, "Pandagpt: One model to instruction-follow them all," *arXiv preprint arXiv:2305.16355*, 2023.

[415] C. Sun, A. Shrivastava, S. Singh, and A. Gupta, "Revisiting unreasonable effectiveness of data in deep learning era," in *ICCV*, 2017.

[416] Q. Sun, Q. Yu, Y. Cui, F. Zhang, X. Zhang, Y. Wang, H. Gao, J. Liu, T. Huang, and X. Wang, "Generative pretraining in multimodality," *arXiv preprint arXiv:2307.05222*, 2023.

[417] Y. Sun, Y. Yang, H. Peng, Y. Shen, Y. Yang, H. Hu, L. Qiu, and H. Koike, "Imagebrush: Learning visual in-context instructions for exemplar-based image manipulation," *arXiv preprint arXiv:2308.00906*, 2023.

[418] Y. Sun, C. Zhu, S. Zheng, K. Zhang, Z. Shui, X. Yu, Y. Zhao, H. Li, Y. Zhang, R. Zhao, *et al.*, "Pathasst: Redefining pathology through generative foundation ai assistant for pathology," *arXiv preprint arXiv:2305.15072*, 2023.

[419] D. Surís, S. Menon, and C. Vondrick, "Vipergpt: Visual inference via python execution for reasoning," *arXiv preprint arXiv:2303.08128*, 2023.

[420] *Svd-xt*, 2023. URL: https://huggingface.co/stabilityai/stable-video-diffusion-img2vid-xt.

[421] H. Tan and M. Bansal, "LXMERT: Learning cross-modality encoder representations from transformers," in *EMNLP*, 2019.

[422] L. Tang, H. Xiao, and B. Li, "Can sam segment anything? when sam meets camouflaged object detection," *arXiv preprint arXiv:2304.04709*, 2023.

[423] Z. Tang, Z. Yang, C. Zhu, M. Zeng, and M. Bansal, "Any-to-any generation via composable diffusion," 2023. arXiv: 2305.11846 [cs.CV].

[424] C. Tao, X. Zhu, W. Su, G. Huang, B. Li, J. Zhou, Y. Qiao, X. Wang, and J. Dai, "Siamese image modeling for self-supervised vision representation learning," in *CVPR*, 2023.

[425] R. Taori, I. Gulrajani, T. Zhang, Y. Dubois, X. Li, C. Guestrin, P. Liang, and T. B. Hashimoto, *Stanford alpaca: An instruction-following llama model*, 2023. URL: https://github.com/tatsu-lab/stanford_alpaca.

[426] G. G. Team, *Gemini: A family of highly capable multimodal models*, 2023. arXiv: 2312.11805 [cs.CL].

[427] The Vicuna Team, *Vicuna: An open-source chatbot impressing GPT-4 with 90%* ChatGPT quality*, 2023. URL: https://vicuna.lmsys.org/.

[428] B. Thomee, D. A. Shamma, G. Friedland, B. Elizalde, K. Ni, D. Poland, D. Borth, and L.-J. Li, "Yfcc100m: The new data in multimedia research," *Communications of the ACM*, 2016.

[429] Y. Tian, D. Krishnan, and P. Isola, "Contrastive multiview coding," in *ECCV*, 2020.

[430] Z. Tian, C. Shen, and H. Chen, "Conditional convolutions for instance segmentation," in *Computer Vision–ECCV 2020: 16th European Conference, Glasgow, UK, August 23–28, 2020, Proceedings, Part I 16*, Springer, pp. 282–298, 2020.

[431] Z. Tong, Y. Song, J. Wang, and L. Wang, "Videomae: Masked autoencoders are data-efficient learners for self-supervised video pre-training," *NeurIPS*, 2022.

[432] H. Touvron, M. Cord, M. Douze, F. Massa, A. Sablayrolles, and H. Jégou, "Training data-efficient image transformers & distillation through attention," in *ICML*, 2021.

[433] H. Touvron, T. Lavril, G. Izacard, X. Martinet, M.-A. Lachaux, T. Lacroix, B. Rozière, N. Goyal, E. Hambro, F. Azhar, *et al.*, "Llama: Open and efficient foundation language models," *arXiv preprint arXiv:2302.13971*, 2023.

[434] M. Tschannen, M. Kumar, A. Steiner, X. Zhai, N. Houlsby, and L. Beyer, "Image captioners are scalable vision learners too," *arXiv preprint arXiv:2306.07915*, 2023.

[435] T. Tu, S. Azizi, D. Driess, M. Schaekermann, M. Amin, P.-C. Chang, A. Carroll, C. Lau, R. Tanno, I. Ktena, *et al.*, "Towards generalist biomedical ai," *arXiv preprint arXiv:2307.14334*, 2023.

[436] A. Vahdat and J. Kautz, "Nvae: A deep hierarchical variational autoencoder," 2020.

[437] A. Vaswani, N. Shazeer, N. Parmar, J. Uszkoreit, L. Jones, A. N. Gomez, Ł. Kaiser, and I. Polosukhin, "Attention is all you need," in *NeurIPS*, 2017.

[438] O. Vinyals, A. Toshev, S. Bengio, and D. Erhan, "Show and tell: A neural image caption generator," in *CVPR*, 2015.

[439] O. Vinyals, A. Toshev, S. Bengio, and D. Erhan, "Show and tell: Lessons learned from the 2015 mscoco image captioning challenge," *IEEE transactions on pattern analysis and machine intelligence*, vol. 39, no. 4, 2016, pp. 652–663.

[440] P. Viola and M. Jones, "Rapid object detection using a boosted cascade of simple features," in *Proceedings of the 2001 IEEE computer society conference on computer vision and pattern recognition. CVPR 2001*, Ieee, vol. 1, pp. I–I, 2001.

[441] B. Wang and A. Komatsuzaki, *GPT-J-6B: A 6 Billion Parameter Autoregressive Language Model*, May 2021. URL: https://github.com/kingoflolz/mesh-transformer-jax.

[442] B. Wang, F. Wu, X. Han, J. Peng, H. Zhong, P. Zhang, X. Dong, W. Li, W. Li, J. Wang, *et al.*, "Vigc: Visual instruction generation and correction," *arXiv preprint arXiv:2308.12714*, 2023.

[443] F. Wang, T. Kong, R. Zhang, H. Liu, and H. Li, "Self-supervised learning by estimating twin class distribution," *TIP*, 2023.

[444] H. Wang, Y. Wang, Z. Zhou, X. Ji, Z. Li, D. Gong, J. Zhou, and W. Liu, "Cosface: Large margin cosine loss for deep face recognition," *CVPR*, 2018.

[445] H. Wang, Y. Zhu, H. Adam, A. Yuille, and L.-C. Chen, "Max-deeplab: End-to-end panoptic segmentation with mask transformers," in *Proceedings of the IEEE/CVF Conference on Computer Vision and Pattern Recognition*, pp. 5463–5474, 2021.

[446] J. Wang, Z. Yang, X. Hu, L. Li, K. Lin, Z. Gan, Z. Liu, C. Liu, and L. Wang, "Git: A generative image-to-text transformer for vision and language," *arXiv preprint arXiv:2205.14100*, 2022.

[447] J. Wang, D. Chen, C. Luo, X. Dai, L. Yuan, Z. Wu, and Y.-G. Jiang, "Chatvideo: A tracklet-centric multimodal and versatile video understanding system," *arXiv preprint arXiv:2304.14407*, 2023.

[448] J. Wang, L. Meng, Z. Weng, B. He, Z. Wu, and Y.-G. Jiang, *To see is to believe: Prompting gpt-4v for better visual instruction tuning*, 2023. arXiv: 2311.07574 [cs.CV].

[449] J. Wang, Y. Zhou, G. Xu, P. Shi, C. Zhao, H. Xu, Q. Ye, M. Yan, J. Zhang, J. Zhu, *et al.*, "Evaluation and analysis of hallucination in large vision-language models," *arXiv preprint arXiv:2308.15126*, 2023.

[450] L. Wang, B. Huang, Z. Zhao, Z. Tong, Y. He, Y. Wang, Y. Wang, and Y. Qiao, "Videomae v2: Scaling video masked autoencoders with dual masking," in *CVPR*, 2023.

[451] P. Wang, A. Yang, R. Men, J. Lin, S. Bai, Z. Li, J. Ma, C. Zhou, J. Zhou, and H. Yang, "Ofa: Unifying architectures, tasks, and modalities through a simple sequence-to-sequence learning framework," in *ICML*, 2022.

[452] R. Wang, D. Chen, Z. Wu, Y. Chen, X. Dai, M. Liu, Y.-G. Jiang, L. Zhou, and L. Yuan, "Bevt: Bert pretraining of video transformers," in *Proceedings of the IEEE/CVF conference on computer vision and pattern recognition*, pp. 14 733–14 743, 2022.

[453] T. Wang, L. Li, K. Lin, C.-C. Lin, Z. Yang, H. Zhang, Z. Liu, and L. Wang, "Disco: Disentangled control for referring human dance generation in real world," *arXiv preprint arXiv:2307.00040*, 2023.

[454] T. Wang, J. Zhang, J. Fei, Y. Ge, H. Zheng, Y. Tang, Z. Li, M. Gao, S. Zhao, Y. Shan, *et al.*, "Caption anything: Interactive image description with diverse multimodal controls," *arXiv preprint arXiv:2305.02677*, 2023.

[455] W. Wang, J. Liu, Z. Lin, J. Yan, S. Chen, C. Low, T. Hoang, J. Wu, J. H. Liew, H. Yan, D. Zhou, and J. Feng, *Magicvideo-v2: Multi-stage high-aesthetic video generation*, 2024. arXiv: 2401.04468 [cs.CV].

[456] W. Wang, Z. Chen, X. Chen, J. Wu, X. Zhu, G. Zeng, P. Luo, T. Lu, J. Zhou, Y. Qiao, *et al.*, "VisionLLM: Large language model is also an open-ended decoder for vision-centric tasks," *arXiv preprint arXiv:2305.11175*, 2023.

[457] W. Wang, H. Bao, L. Dong, J. Bjorck, Z. Peng, Q. Liu, K. Aggarwal, O. K. Mohammed, S. Singhal, S. Som, and F. Wei, "Image as a foreign language: Beit pretraining for all vision and vision-language tasks," *arXiv preprint arXiv:2208.10442*, 2022.

[458] W. Wang, H. Bao, L. Dong, and F. Wei, "Vlmo: Unified vision-language pre-training with mixture-of-modality-experts," *arXiv preprint arXiv:2111.02358*, 2021.

[459] X. Wang, W. Wang, Y. Cao, C. Shen, and T. Huang, "Images speak in images: A generalist painter for in-context visual learning," in *Proceedings of the IEEE/CVF Conference on Computer Vision and Pattern Recognition*, pp. 6830–6839, 2023.

[460] X. Wang, X. Zhang, Y. Cao, W. Wang, C. Shen, and T. Huang, "Seggpt: Segmenting everything in context," *arXiv preprint arXiv:2304.03284*, 2023.

[461] Y. Wang, H. Ivison, P. Dasigi, J. Hessel, T. Khot, K. R. Chandu, D. Wadden, K. MacMillan, N. A. Smith, I. Beltagy, *et al.*, "How far can camels go? exploring the state of instruction tuning on open resources," *arXiv preprint arXiv:2306.04751*, 2023.

[462] Y. Wang, Y. Kordi, S. Mishra, A. Liu, N. A. Smith, D. Khashabi, and H. Hajishirzi, "Self-instruct: Aligning language model with self generated instructions," *arXiv preprint arXiv:2212.10560*, 2022.

[463] Y. Wang, S. Mishra, P. Alipoormolabashi, Y. Kordi, A. Mirzaei, A. Arunkumar, A. Ashok, A. S. Dhanasekaran, A. Naik, D. Stap, *et al.*, "Benchmarking generalization via in-context instructions on 1,600+ language tasks," *arXiv preprint arXiv:2204.07705*, 2022.

[464] Z. Wang, H. Huang, Y. Zhao, Z. Zhang, and Z. Zhao, "Chat-3d: Data-efficiently tuning large language model for universal dialogue of 3d scenes," *arXiv preprint arXiv:2308.08769*, 2023.

[465] Z. Wang, Y. Jiang, Y. Lu, Y. Shen, P. He, W. Chen, Z. Wang, and M. Zhou, *In-context learning unlocked for diffusion models*, 2023. arXiv: 2305.01115 [cs.CV].

[466] Z. Wang, C. Lu, Y. Wang, F. Bao, C. Li, H. Su, and J. Zhu, "Prolificdreamer: High-fidelity and diverse text-to-3d generation with variational score distillation," *arXiv preprint arXiv:2305.16213*, 2023.

[467] Z. Wang, J. Chen, and S. C. Hoi, "Deep learning for image super-resolution: A survey," *IEEE transactions on pattern analysis and machine intelligence*, vol. 43, no. 10, 2020, pp. 3365–3387.

[468] Z. Wang, J. Yu, A. W. Yu, Z. Dai, Y. Tsvetkov, and Y. Cao, "Simvlm: Simple visual language model pretraining with weak supervision," in *ICLR*, 2022.

[469] F. Weers, V. Shankar, A. Katharopoulos, Y. Yang, and T. Gunter, "Masked autoencoding does not help natural language supervision at scale," in *CVPR*, 2023.

[470] C. Wei, H. Fan, S. Xie, C.-Y. Wu, A. Yuille, and C. Feichtenhofer, "Masked feature prediction for self-supervised visual pre-training," in *CVPR*, 2021.

[471] J. Wei, X. Wang, D. Schuurmans, M. Bosma, E. Chi, Q. Le, and D. Zhou, "Chain of thought prompting elicits reasoning in large language models," *arXiv preprint arXiv:2201.11903*, 2022.

[472] L. Wei, Z. Jiang, W. Huang, and L. Sun, "Instructiongpt-4: A 200-instruction paradigm for fine-tuning minigpt-4," *arXiv preprint arXiv:2308.12067*, 2023.

[473] L. Wei, L. Xie, W. Zhou, H. Li, and Q. Tian, "Mvp: Multimodality-guided visual pre-training," in *ECCV*, 2022.

[474] Y. Wei, Y. Zhang, Z. Ji, J. Bai, L. Zhang, and W. Zuo, "Elite: Encoding visual concepts into textual embeddings for customized text-to-image generation," *arXiv preprint arXiv:2302.13848*, 2023.

[475] L. Weng, "Llm-powered autonomous agents," *lilianweng.github.io*, Jun. 2023. URL: https://lilianweng.github.io/posts/2023-06-23-agent/.

[476] C. Wu, X. Zhang, Y. Zhang, Y. Wang, and W. Xie, "Towards generalist foundation model for radiology," *arXiv preprint arXiv:2308.02463*, 2023.

[477] C. Wu, S. Yin, W. Qi, X. Wang, Z. Tang, and N. Duan, "Visual chatgpt: Talking, drawing and editing with visual foundation models," *arXiv preprint arXiv:2303.04671*, 2023.

[478] J. Wu, J. Wang, Z. Yang, Z. Gan, Z. Liu, J. Yuan, and L. Wang, "Grit: A generative region-to-text transformer for object understanding," *arXiv preprint arXiv:2212.00280*, 2022.

[479] J. Wu, J. Lu, A. Sabharwal, and R. Mottaghi, "Multi-modal answer validation for knowledge-based VQA," *arXiv preprint arXiv:2103.12248*, 2021.

[480] J. Wu, Y. Jiang, P. Sun, Z. Yuan, and P. Luo, "Language as queries for referring video object segmentation," in *Proceedings of the IEEE/CVF Conference on Computer Vision and Pattern Recognition*, pp. 4974–4984, 2022.

[481] S. Wu, H. Fei, L. Qu, W. Ji, and T.-S. Chua, "Next-gpt: Any-to-any multimodal llm," *CoRR*, vol. abs/2309.05519, 2023.

[482] W. Wu, A. Timofeev, C. Chen, B. Zhang, K. Duan, S. Liu, Y. Zheng, J. Shlens, X. Du, Z. Gan, *et al.*, "Mofi: Learning image representations from noisy entity annotated images," *arXiv preprint arXiv:2306.07952*, 2023.

[483] Y. Wu, J. Lim, and M.-H. Yang, "Online object tracking: A benchmark," in *Proceedings of the IEEE conference on computer vision and pattern recognition*, pp. 2411–2418, 2013.

[484] Z. Wu, Y. Xiong, S. X. Yu, and D. Lin, "Unsupervised feature learning via non-parametric instance discrimination," in *CVPR*, 2018.

[485] Y. Xian, C. H. Lampert, B. Schiele, and Z. Akata, "Zero-shot learning—a comprehensive evaluation of the good, the bad and the ugly," *TPAMI*, 2018.

[486] Z. Xiao, Y. Chen, L. Zhang, J. Yao, Z. Wu, X. Yu, Y. Pan, L. Zhao, C. Ma, X. Liu, *et al.*, "Instruction-vit: Multi-modal prompts for instruction learning in vit," *arXiv preprint arXiv:2305.00201*, 2023.

[487] D. Xie, R. Wang, J. Ma, C. Chen, H. Lu, D. Yang, F. Shi, and X. Lin, "Edit everything: A text-guided generative system for images editing," *arXiv preprint arXiv:2304.14006*, 2023.

[488] X. Xie, L. Fu, Z. Zhang, Z. Wang, and X. Bai, *Toward understanding wordart: Corner-guided transformer for scene text recognition*, 2022. arXiv: 2208.00438 [cs.CV].

[489] Z. Xie, Y. Lin, Z. Yao, Z. Zhang, Q. Dai, Y. Cao, and H. Hu, "Self-supervised learning with swin transformers," *arXiv preprint arXiv:2105.04553*, 2021.

[490] Z. Xie, Z. Zhang, Y. Cao, Y. Lin, J. Bao, Z. Yao, Q. Dai, and H. Hu, "Simmim: A simple framework for masked image modeling," in *CVPR*, 2022.

[491] Z. Xie, Z. Zhang, Y. Cao, Y. Lin, Y. Wei, Q. Dai, and H. Hu, "On data scaling in masked image modeling," in *CVPR*, 2023.

[492] D. Xu, W. Ouyang, X. Wang, and N. Sebe, "Pad-net: Multi-tasks guided prediction-and-distillation network for simultaneous depth estimation and scene parsing," in *Proceedings of the IEEE Conference on Computer Vision and Pattern Recognition*, pp. 675–684, 2018.

[493] H. Xu, M. Yan, C. Li, B. Bi, S. Huang, W. Xiao, and F. Huang, *E2e-vlp: End-to-end vision-language pre-training enhanced by visual learning*, 2021. arXiv: 2106.01804 [cs.CV].

[494] J. Xu, S. De Mello, S. Liu, W. Byeon, T. Breuel, J. Kautz, and X. Wang, "Groupvit: Semantic segmentation emerges from text supervision," in *CVPR*, 2022.

[495] J. Xu, S. Liu, A. Vahdat, W. Byeon, X. Wang, and S. De Mello, "Open-vocabulary panoptic segmentation with text-to-image diffusion models," in *Proceedings of the IEEE/CVF Conference on Computer Vision and Pattern Recognition*, pp. 2955–2966, 2023.

[496] J. Xu, L. Xu, Y. Yang, X. Li, Y. Xie, Y.-J. Huang, and Y. Li, "U-llava: Unifying multi-modal tasks via large language model," *arXiv preprint arXiv:2311.05348*, 2023.

[497] P. Xu, W. Shao, K. Zhang, P. Gao, S. Liu, M. Lei, F. Meng, S. Huang, Y. Qiao, and P. Luo, "Lvlm-ehub: A comprehensive evaluation benchmark for large vision-language models," *arXiv preprint arXiv:2306.09265*, 2023.

[498] R. Xu, X. Wang, T. Wang, Y. Chen, J. Pang, and D. Lin, "Pointllm: Empowering large language models to understand point clouds," *arXiv preprint arXiv:2308.16911*, 2023.

[499] Y. Xu, Y. Zhao, Z. Xiao, and T. Hou, *Ufogen: You forward once large scale text-to-image generation via diffusion gans*, 2023. arXiv: 2311.09257 [cs.CV].

[500] Z. Xu, Y. Shen, and L. Huang, "Multiinstruct: Improving multi-modal zero-shot learning via instruction tuning," *arXiv preprint arXiv:2212.10773*, 2022.

[501] B. Yan, Y. Jiang, J. Wu, D. Wang, P. Luo, Z. Yuan, and H. Lu, "Universal instance perception as object discovery and retrieval," in *Proceedings of the IEEE/CVF Conference on Computer Vision and Pattern Recognition*, pp. 15 325–15 336, 2023.

[502] B. Yang, S. Gu, B. Zhang, T. Zhang, X. Chen, X. Sun, D. Chen, and F. Wen, "Paint by example: Exemplar-based image editing with diffusion models," in *Proceedings of the IEEE/CVF Conference on Computer Vision and Pattern Recognition*, pp. 18 381–18 391, 2023.

[503] J. Yang, C. Li, P. Zhang, B. Xiao, C. Liu, L. Yuan, and J. Gao, "Unified contrastive learning in image-text-label space," in *CVPR*, 2022.

[504] J. Yang, C. Li, P. Zhang, B. Xiao, L. Yuan, C. Liu, and J. Gao, "Unicl: Unified contrastive learning in image-text-label space," in *CVPR*, 2022.

[505] J. Yang, M. Gao, Z. Li, S. Gao, F. Wang, and F. Zheng, "Track anything: Segment anything meets videos," *arXiv preprint arXiv:2304.11968*, 2023.

[506] R. Yang, L. Song, Y. Li, S. Zhao, Y. Ge, X. Li, and Y. Shan, "Gpt4tools: Teaching large language model to use tools via self-instruction," *arXiv preprint arXiv:2305.18752*, 2023.

[507] Z. Yang, J. Wang, Y. Tang, K. Chen, H. Zhao, and P. H. Torr, "Lavt: Language-aware vision transformer for referring image segmentation," in *Proceedings of the IEEE/CVF Conference on Computer Vision and Pattern Recognition*, pp. 18 155–18 165, 2022.

[508] Z. Yang, Z. Gan, J. Wang, X. Hu, F. Ahmed, Z. Liu, Y. Lu, and L. Wang, "Crossing the format boundary of text and boxes: Towards unified vision-language modeling," *arXiv preprint arXiv:2111.12085*, 2021.

[509] Z. Yang, Z. Gan, J. Wang, X. Hu, F. Ahmed, Z. Liu, Y. Lu, and L. Wang, "Unitab: Unifying text and box outputs for grounded vision-language modeling," in *European Conference on Computer Vision*, Springer, pp. 521–539, 2022.

[510] Z. Yang, Z. Gan, J. Wang, X. Hu, Y. Lu, Z. Liu, and L. Wang, "An empirical study of gpt-3 for few-shot knowledge-based vqa," in *AAAI*, 2022.

[511] Z. Yang, J. Wang, Z. Gan, L. Li, K. Lin, C. Wu, N. Duan, Z. Liu, C. Liu, M. Zeng, *et al.*, "Reco: Region-controlled text-to-image generation," in *Proceedings of the IEEE/CVF Conference on Computer Vision and Pattern Recognition*, pp. 14 246–14 255, 2023.

[512] Z. Yang, W. Ping, Z. Liu, V. Korthikanti, W. Nie, D.-A. Huang, L. Fan, Z. Yu, S. Lan, B. Li, *et al.*, "Re-vilm: Retrieval-augmented visual language model for zero and few-shot image captioning," *arXiv preprint arXiv:2302.04858*, 2023.

[513] Z. Yang*, L. Li*, J. Wang*, K. Lin*, E. Azarnasab*, F. Ahmed*, Z. Liu, C. Liu, M. Zeng, and L. Wang, "Mm-react: Prompting chatgpt for multimodal reasoning and action," 2023.

[514] L. Yao, J. Han, X. Liang, D. Xu, W. Zhang, Z. Li, and H. Xu, "Detclipv2: Scalable open-vocabulary object detection pre-training via word-region alignment," in *Proceedings of the IEEE/CVF Conference on Computer Vision and Pattern Recognition*, pp. 23 497–23 506, 2023.

[515] L. Yao, J. Han, Y. Wen, X. Liang, D. Xu, W. Zhang, Z. Li, C. Xu, and H. Xu, "Detclip: Dictionary-enriched visual-concept paralleled pre-training for open-world detection," *arXiv preprint arXiv:2209.09407*, 2022.

[516] L. Yao, R. Huang, L. Hou, G. Lu, M. Niu, H. Xu, X. Liang, Z. Li, X. Jiang, and C. Xu, "Filip: Fine-grained interactive language-image pre-training," in *ICLR*, 2022.

[517] S. Yao, J. Zhao, D. Yu, N. Du, I. Shafran, K. Narasimhan, and Y. Cao, "React: Synergizing reasoning and acting in language models," *arXiv preprint arXiv:2210.03629*, 2022.

[518] M. Yasunaga, A. Aghajanyan, W. Shi, R. James, J. Leskovec, P. Liang, M. Lewis, L. Zettlemoyer, and W.-t. Yih, "Retrieval-augmented multimodal language modeling," *arXiv preprint arXiv:2211.12561*, 2022.

[519] J. Ye, A. Hu, H. Xu, Q. Ye, M. Yan, Y. Dan, C. Zhao, G. Xu, C. Li, J. Tian, *et al.*, "Mplug-docowl: Modularized multimodal large language model for document understanding," *arXiv preprint arXiv:2307.02499*, 2023.

[520] L. Ye, M. Rochan, Z. Liu, and Y. Wang, "Cross-modal self-attention network for referring image segmentation," in *Proceedings of the IEEE/CVF conference on computer vision and pattern recognition*, pp. 10 502–10 511, 2019.

[521] M. Ye, X. Zhang, P. C. Yuen, and S.-F. Chang, "Unsupervised embedding learning via invariant and spreading instance feature," in *CVPR*, 2019.

[522] Q. Ye, H. Xu, G. Xu, J. Ye, M. Yan, Y. Zhou, J. Wang, A. Hu, P. Shi, Y. Shi, *et al.*, "Mplug-owl: Modularization empowers large language models with multimodality," *arXiv preprint arXiv:2304.14178*, 2023.

[523] K. Yi, Y. Ge, X. Li, S. Yang, D. Li, J. Wu, Y. Shan, and X. Qie, "Masked image modeling with denoising contrast," *arXiv preprint arXiv:2205.09616*, 2022.

[524] A. Yilmaz, O. Javed, and M. Shah, "Object tracking: A survey," *Acm computing surveys (CSUR)*, vol. 38, no. 4, 2006, 13–es.

[525] D. Yin, L. Dong, H. Cheng, X. Liu, K.-W. Chang, F. Wei, and J. Gao, "A survey of knowledge-intensive nlp with pre-trained language models," *arXiv preprint arXiv:2202.08772*, 2022.

[526] Z. Yin, J. Wang, J. Cao, Z. Shi, D. Liu, M. Li, L. Sheng, L. Bai, X. Huang, Z. Wang, *et al.*, "Lamm: Language-assisted multi-modal instruction-tuning dataset, framework, and benchmark," *arXiv preprint arXiv:2306.06687*, 2023.

[527] S. Yoon, W. Y. Kang, S. Jeon, S. Lee, C. Han, J. Park, and E.-S. Kim, "Image-to-image retrieval by learning similarity between scene graphs," in *Proceedings of the AAAI Conference on Artificial Intelligence*, vol. 35, pp. 10 718–10 726, 2021.

[528] H. You, H. Zhang, Z. Gan, X. Du, B. Zhang, Z. Wang, L. Cao, S.-F. Chang, and Y. Yang, "Ferret: Refer and ground anything anywhere at any granularity," *arXiv preprint arXiv:2310.07704*, 2023.

[529] J. Yu, X. Li, J. Y. Koh, H. Zhang, R. Pang, J. Qin, A. Ku, Y. Xu, J. Baldridge, and Y. Wu, "Vector-quantized image modeling with improved vqgan," *arXiv preprint arXiv:2110.04627*, 2021.

[530] J. Yu, Z. Wang, V. Vasudevan, L. Yeung, M. Seyedhosseini, and Y. Wu, "Coca: Contrastive captioners are image-text foundation models," *TMLR*, 2022.

[531] J. Yu, Y. Xu, J. Y. Koh, T. Luong, G. Baid, Z. Wang, V. Vasudevan, A. Ku, Y. Yang, B. K. Ayan, *et al.*, "Scaling autoregressive models for content-rich text-to-image generation," *Transactions on Machine Learning Research*, 2022.

[532] L. Yu, P. Poirson, S. Yang, A. C. Berg, and T. L. Berg, "Modeling context in referring expressions," in *ECCV*, 2016.

[533] L. Yu and et al, "Scaling autoregressive multi-modal models: Pretraining and instruction tuning," 2023.

[534] Q. Yu, J. Li, W. Ye, S. Tang, and Y. Zhuang, "Interactive data synthesis for systematic vision adaptation via llms-aigcs collaboration," *arXiv preprint arXiv:2305.12799*, 2023.

[535] Q. Yu, J. He, X. Deng, X. Shen, and L.-C. Chen, "Convolutions die hard: Open-vocabulary segmentation with single frozen convolutional clip," *arXiv preprint arXiv:2308.02487*, 2023.

[536] T. Yu, R. Feng, R. Feng, J. Liu, X. Jin, W. Zeng, and Z. Chen, "Inpaint anything: Segment anything meets image inpainting," *arXiv preprint arXiv:2304.06790*, 2023.

[537] W. Yu, Z. Yang, L. Li, J. Wang, K. Lin, Z. Liu, X. Wang, and L. Wang, "Mm-vet: Evaluating large multimodal models for integrated capabilities," *arXiv preprint arXiv:2308.02490*, 2023.

[538] Z. Yu, J. Yu, Y. Cui, D. Tao, and Q. Tian, "Deep modular co-attention networks for visual question answering," in *CVPR*, 2019.

[539] L. Yuan, D. Chen, Y.-L. Chen, N. Codella, X. Dai, J. Gao, H. Hu, X. Huang, B. Li, C. Li, C. Liu, M. Liu, Z. Liu, Y. Lu, Y. Shi, L. Wang, J. Wang, B. Xiao, Z. Xiao, J. Yang, M. Zeng, L. Zhou, and P. Zhang, "Florence: A new foundation model for computer vision," *arXiv preprint arXiv:2111.11432*, 2021.

[540] A. R. Zamir, A. Sax, W. Shen, L. J. Guibas, J. Malik, and S. Savarese, "Taskonomy: Disentangling task transfer learning," in *Proceedings of the IEEE conference on computer vision and pattern recognition*, pp. 3712–3722, 2018.

[541] Y. Zang, W. Li, J. Han, K. Zhou, and C. C. Loy, "Contextual object detection with multimodal large language models," *arXiv preprint arXiv:2305.18279*, 2023.

[542] Y. Zang, W. Li, K. Zhou, C. Huang, and C. C. Loy, "Open-vocabulary detr with conditional matching," *arXiv preprint arXiv:2203.11876*, 2022.

[543] A. Zareian, K. D. Rosa, D. H. Hu, and S.-F. Chang, "Open-vocabulary object detection using captions," in *CVPR*, 2021.

[544] J. Zbontar, L. Jing, I. Misra, Y. LeCun, and S. Deny, "Barlow twins: Self-supervised learning via redundancy reduction," in *ICML*, 2021.

[545] R. Zellers, Y. Bisk, A. Farhadi, and Y. Choi, "From recognition to cognition: Visual commonsense reasoning," in *CVPR*, 2019.

[546] Y. Zeng, X. Zhang, and H. Li, "Multi-grained vision language pre-training: Aligning texts with visual concepts," in *ICML*, 2022.

[547] Y. Zeng, Z. Lin, J. Zhang, Q. Liu, J. Collomosse, J. Kuen, and V. M. Patel, "Scenecomposer: Any-level semantic image synthesis," in *Proceedings of the IEEE/CVF Conference on Computer Vision and Pattern Recognition*, pp. 22 468–22 478, 2023.

[548] X. Zhai, A. Kolesnikov, N. Houlsby, and L. Beyer, "Scaling vision transformers," in *CVPR*, 2022.

[549] X. Zhai, B. Mustafa, A. Kolesnikov, and L. Beyer, "Sigmoid loss for language image pre-training," *arXiv preprint arXiv:2303.15343*, 2023.

[550] X. Zhai, X. Wang, B. Mustafa, A. Steiner, D. Keysers, A. Kolesnikov, and L. Beyer, "Lit: Zero-shot transfer with locked-image text tuning," in *CVPR*, 2022.

[551] C. Zhang, Z. Yang, X. He, and L. Deng, "Multimodal intelligence: Representation learning, information fusion, and applications," *JSTSP*, 2020.

[552] C. Zhang, S. Zheng, C. Li, Y. Qiao, T. Kang, X. Shan, C. Zhang, C. Qin, F. Rameau, S.-H. Bae, *et al.*, "A survey on segment anything model (sam): Vision foundation model meets prompt engineering," *arXiv preprint arXiv:2306.06211*, 2023.

[553] C. Zhang, C. Zhang, M. Zhang, and I. S. Kweon, "Text-to-image diffusion model in generative ai: A survey," *arXiv preprint arXiv:2303.07909*, 2023.

[554] C. Zhang, L. Liu, Y. Cui, G. Huang, W. Lin, Y. Yang, and Y. Hu, "A comprehensive survey on segment anything model for vision and beyond," *arXiv preprint arXiv:2305.08196*, 2023.

[555] H. Zhang, X. Li, and L. Bing, "Video-llama: An instruction-tuned audio-visual language model for video understanding," *arXiv preprint arXiv:2306.02858*, 2023.

[556] H. Zhang, F. Li, S. Liu, L. Zhang, H. Su, J. Zhu, L. M. Ni, and H.-Y. Shum, "Dino: Detr with improved denoising anchor boxes for end-to-end object detection," *arXiv preprint arXiv:2203.03605*, 2022.

[557] H. Zhang, F. Li, X. Zou, S. Liu, C. Li, J. Gao, J. Yang, and L. Zhang, "A simple framework for open-vocabulary segmentation and detection," *arXiv preprint arXiv:2303.08131*, 2023.

[558] H. Zhang, H. Li, F. Li, T. Ren, X. Zou, S. Liu, S. Huang, J. Gao, L. Zhang, C. Li, and J. Yang, *Llava-grounding: Grounded visual chat with large multimodal models*, 2023.

[559] H. Zhang, P. Zhang, X. Hu, Y.-C. Chen, L. H. Li, X. Dai, L. Wang, L. Yuan, J.-N. Hwang, and J. Gao, "Glipv2: Unifying localization and vision-language understanding," in *ECCV*, 2022.

[560] J. Zhang, J. Huang, S. Jin, and S. Lu, "Vision-language models for vision tasks: A survey," *arXiv preprint arXiv:2304.00685*, 2023.

[561] J. Zhang, F. Shen, L. Liu, F. Zhu, M. Yu, L. Shao, H. T. Shen, and L. Van Gool, "Generative domain-migration hashing for sketch-to-image retrieval," in *Proceedings of the European conference on computer vision (ECCV)*, pp. 297–314, 2018.

[562] L. Zhang and M. Agrawala, "Adding conditional control to text-to-image diffusion models," *arXiv preprint arXiv:2302.05543*, 2023.

[563] P. Zhang, X. Li, X. Hu, J. Yang, L. Zhang, L. Wang, Y. Choi, and J. Gao, "VinVL: Revisiting visual representations in vision-language models," in *CVPR*, 2021.

[564] R. Zhang, Z. Jiang, Z. Guo, S. Yan, J. Pan, H. Dong, P. Gao, and H. Li, "Personalize segment anything model with one shot," *arXiv preprint arXiv:2305.03048*, 2023.

[565] S. Zhang, L. Dong, X. Li, S. Zhang, X. Sun, S. Wang, J. Li, R. Hu, T. Zhang, F. Wu, *et al.*, "Instruction tuning for large language models: A survey," *arXiv preprint arXiv:2308.10792*, 2023.

[566] S. Zhang, P. Sun, S. Chen, M. Xiao, W. Shao, W. Zhang, K. Chen, and P. Luo, "Gpt4roi: Instruction tuning large language model on region-of-interest," *arXiv preprint arXiv:2307.03601*, 2023.

[567] S. Zhang, C. Gong, L. Wu, X. Liu, and M. Zhou, "Automl-gpt: Automatic machine learning with gpt," *arXiv preprint arXiv:2305.02499*, 2023.

[568] T. Zhang, V. Kishore, F. Wu, K. Q. Weinberger, and Y. Artzi, "Bertscore: Evaluating text generation with bert," *arXiv preprint arXiv:1904.09675*, 2019.

[569] W. Zhang, S. M. Aljunied, C. Gao, Y. K. Chia, and L. Bing, "M3exam: A multilingual, multimodal, multilevel benchmark for examining large language models," *arXiv preprint arXiv:2306.05179*, 2023.

[570] X. Zhang, C. Wu, Z. Zhao, W. Lin, Y. Zhang, Y. Wang, and W. Xie, "Pmc-vqa: Visual instruction tuning for medical visual question answering," *arXiv preprint arXiv:2305.10415*, 2023.

[571] X. Zhang, Y. Tian, W. Huang, Q. Ye, Q. Dai, L. Xie, and Q. Tian, "Hivit: Hierarchical vision transformer meets masked image modeling," *arXiv preprint arXiv:2205.14949*, 2022.

[572] X. Zhang, J. Chen, J. Yuan, Q. Chen, J. Wang, X. Wang, S. Han, X. Chen, J. Pi, K. Yao, *et al.*, "Cae v2: Context autoencoder with clip target," *arXiv preprint arXiv:2211.09799*, 2022.

[573] Y. Zhang, R. Zhang, J. Gu, Y. Zhou, N. Lipka, D. Yang, and T. Sun, "Llavar: Enhanced visual instruction tuning for text-rich image understanding," *arXiv preprint arXiv:2306.17107*, 2023.

[574] Y. Zhang and R. Jiao, "How segment anything model (sam) boost medical image segmentation?" *arXiv preprint arXiv:2305.03678*, 2023.

[575] Y. Zhang, X. Huang, J. Ma, Z. Li, Z. Luo, Y. Xie, Y. Qin, T. Luo, Y. Li, S. Liu, *et al.*, "Recognize anything: A strong image tagging model," *arXiv preprint arXiv:2306.03514*, 2023.

[576] B. Zhao, B. Wu, and T. Huang, "Svit: Scaling up visual instruction tuning," *arXiv preprint arXiv:2307.04087*, 2023.
[577] S. Zhao, D. Chen, Y.-C. Chen, J. Bao, S. Hao, L. Yuan, and K.-Y. K. Wong, "Uni-controlnet: All-in-one control to text-to-image diffusion models," *arXiv preprint arXiv:2305.16322*, 2023.
[578] Y. Zhao, Z. Lin, D. Zhou, Z. Huang, J. Feng, and B. Kang, "Bubogpt: Enabling visual grounding in multi-modal llms," *arXiv preprint arXiv:2307.08581*, 2023.
[579] Y. Zhao, T. Pang, C. Du, X. Yang, C. Li, N.-M. Cheung, and M. Lin, "On evaluating adversarial robustness of large vision-language models," *arXiv preprint arXiv:2305.16934*, 2023.
[580] Z. Zhao, E. Wallace, S. Feng, D. Klein, and S. Singh, "Calibrate before use: Improving few-shot performance of language models," in *International Conference on Machine Learning*, PMLR, pp. 12 697–12 706, 2021.
[581] Y. Zhong, J. Yang, P. Zhang, C. Li, N. Codella, L. H. Li, L. Zhou, X. Dai, L. Yuan, Y. Li, *et al.*, "Regionclip: Region-based language-image pretraining," in *CVPR*, 2022.
[582] Y. Zhong, J. Yang, P. Zhang, C. Li, N. Codella, L. H. Li, L. Zhou, X. Dai, L. Yuan, Y. Li, *et al.*, "Regionclip: Region-based language-image pretraining," in *Proceedings of the IEEE/CVF Conference on Computer Vision and Pattern Recognition*, pp. 16 793–16 803, 2022.
[583] B. Zhou, H. Zhao, X. Puig, S. Fidler, A. Barriuso, and A. Torralba, "Scene parsing through ade20k dataset," in *Proceedings of the IEEE conference on computer vision and pattern recognition*, pp. 633–641, 2017.
[584] C. Zhou, C. C. Loy, and B. Dai, "Extract free dense labels from clip," in *ECCV*, 2022.
[585] C. Zhou, P. Liu, P. Xu, S. Iyer, J. Sun, Y. Mao, X. Ma, A. Efrat, P. Yu, L. Yu, *et al.*, "Lima: Less is more for alignment," *arXiv preprint arXiv:2305.11206*, 2023.
[586] G. Zhou, Y. Hong, and Q. Wu, "Navgpt: Explicit reasoning in vision-and-language navigation with large language models," *arXiv preprint arXiv:2305.16986*, 2023.

[587] J. Zhou, L. Dong, Z. Gan, L. Wang, and F. Wei, "Non-contrastive learning meets language-image pre-training," in *CVPR*, 2023.

[588] J. Zhou, C. Wei, H. Wang, W. Shen, C. Xie, A. Yuille, and T. Kong, "Ibot: Image bert pre-training with online tokenizer," *arXiv preprint arXiv:2111.07832*, 2021.

[589] T. Zhou, Y. Zhang, Y. Zhou, Y. Wu, and C. Gong, "Can sam segment polyps?" *arXiv preprint arXiv:2304.07583*, 2023.

[590] X. Zhou, R. Girdhar, A. Joulin, P. Krähenbühl, and I. Misra, "Detecting twenty-thousand classes using image-level supervision," in *European Conference on Computer Vision*, Springer, pp. 350–368, 2022.

[591] Y. Zhou, C. Li, C. Chen, J. Gao, and J. Xu, "Lafite2: Few-shot text-to-image generation," *arXiv preprint arXiv:2210.14124*, 2022.

[592] Y. Zhou and N. Shimada, "Vision + language applications: A survey," in *Proceedings of the IEEE/CVF Conference on Computer Vision and Pattern Recognition*, pp. 826–842, 2023.

[593] D. Zhu, J. Chen, X. Shen, X. Li, and M. Elhoseiny, *Minigpt-4: Enhancing vision-language understanding with advanced large language models*, 2023. arXiv: 2304.10592 `[cs.CV]`.

[594] P. Zhu, H. Wang, and V. Saligrama, "Zero shot detection," *IEEE Transactions on Circuits and Systems for Video Technology*, vol. 30, no. 4, 2019, pp. 998–1010.

[595] P. Zhu, H. Wang, and V. Saligrama, "Don't even look once: Synthesizing features for zero-shot detection," in *Proceedings of the IEEE/CVF Conference on Computer Vision and Pattern Recognition*, pp. 11 693–11 702, 2020.

[596] W. Zhu, J. Hessel, A. Awadalla, S. Y. Gadre, J. Dodge, A. Fang, Y. Yu, L. Schmidt, W. Y. Wang, and Y. Choi, "Multimodal c4: An open, billion-scale corpus of images interleaved with text," *arXiv preprint arXiv:2304.06939*, 2023.

[597] Y. Zhu, M. Zhu, N. Liu, Z. Ou, X. Mou, and J. Tang, "Llava-phi: Efficient multi-modal assistant with small language model," *arXiv preprint arXiv:2401.02330*, 2024.

[598] Z. Zong, G. Song, and Y. Liu, *Detrs with collaborative hybrid assignments training*, 2023. arXiv: 2211.12860 `[cs.CV]`.

[599] X. Zou, Z.-Y. Dou, J. Yang, Z. Gan, L. Li, C. Li, X. Dai, H. Behl, J. Wang, L. Yuan, *et al.*, "Generalized decoding for pixel, image, and language," in *Proceedings of the IEEE/CVF Conference on Computer Vision and Pattern Recognition*, 2023.

[600] X. Zou, H. Liu, and Y. J. Lee, "End-to-end instance edge detection," *arXiv preprint arXiv:2204.02898*, 2022.

[601] X. Zou, J. Yang, H. Zhang, F. Li, L. Li, J. Gao, and Y. J. Lee, "Segment everything everywhere all at once," *arXiv preprint arXiv:2304.06718*, 2023.

www.ingramcontent.com/pod-product-compliance
Lightning Source LLC
LaVergne TN
LVHW022017280725
817250LV00016B/323
* 9 7 8 1 6 3 8 2 8 3 3 6 2 *